Debugging at the Electronic System Level

Frank Rogin • Rolf Drechsler

Debugging at the Electronic System Level

Springer

Frank Rogin
Fraunhofer - Institut für
Integrierte Schaltungen
Institutsteil
Entwurfsautomatisierung
Zeunerstr. 38
01069 Dresden
Germany
frank.rogin@web.de

Rolf Drechsler
Universität Bremen
AG Rechnerarchitektur
Bibliothekstr. 1
28359 Bremen
Germany
drechsle@Informatik.Uni-Bremen.de

ISBN 978-90-481-9254-0 e-ISBN 978-90-481-9255-7
DOI 10.1007/978-90-481-9255-7
Springer Dordrecht Heidelberg London New York

Library of Congress Control Number: 2010929863

Cover design: eStudio Calamar S.L.

Printed on acid-free paper

Springer is part of Springer Science+Business Media (www.springer.com)

Contents

List of Figures

List of Tables

Preface

Debugging becomes more and more the bottleneck to chip design productivity, especially while developing modern complex and heterogenous integrated circuits and systems at the *Electronic System Level* (ESL). Today, debugging is still an unsystematic and lengthy process. Here, a simple reporting of a failure is not enough, anymore. Rather, it becomes more and more important not only to find many errors early during development but also to provide efficient methods for their isolation. In this book the state-of-the-art of modeling and verification of ESL designs is reviewed. There, a particular focus is taken onto SystemC. Then, a reasoning hierarchy is introduced. The hierarchy combines well-known debugging techniques with whole new techniques to improve the verification efficiency at ESL. The proposed systematic debugging approach is supported amongst others by static code analysis, debug patterns, dynamic program slicing, design visualization, property generation, and automatic failure isolation. All techniques were empirically evaluated using real-world industrial designs. Summarized, the introduced approach enables a systematic search for errors in ESL designs. Here, the debugging techniques improve and accelerate error detection, observation, and isolation as well as design understanding.

Acknowledgements

We would like to thank all colleagues of Division Design Automation of Fraunhofer Institute for Integrated Circuits in Dresden and all members of the research group for computer architecture in Bremen for the amazing atmosphere during research and work and the many fruitful discussions.

Moreover, we are grateful to all our coauthors of the papers that make up an important part of this book: Thomas Berndt, Erhard Fehlauer, Görschwin Fey, Christian Genz, Christian Haufe, Thomas Klotz, Sebastian Ohnewald, and Steffen Rülke. Additionally, we would like to thank our students who had contributed many prototype implementations: Lars Ehrler, Dirk Linke, Dominic Scharfe, Tom Schiekel, and Richard Wähner.

Erhard Fehlauer, Görschwin Fey, and Steffen Rülke helped us in proofreading and improving the presentation.

FRANK ROGIN AND ROLF DRECHSLER
Bremen, March 2010

Chapter 1

Introduction

Modern integrated circuits and systems consist of many different functional blocks comprising multiple heterogeneous processor cores, dedicated analog/mixed signal components, various on-chip busses and memories, (third-party) *Intellectual Property* (IP), and most notably more and more embedded software. Following "Moore's Law", the available chip capacity grows exponentially. Currently, high-end processor designs reaches up to 2 billion transistors. A complete system can be integrated onto a single chip which is then called *System-on-a-Chip* (SoC). The increasing design complexity and scale of SoC designs combined with non-functional requirements and constraints on the final product, e.g. low power, robustness, reliability, and low cost, make the verification of the design correctness a complex and crucial task. Functional errors are still the most important cause of design re-spins. According to a study from Collett International Research [Cir04] nearly 40% of all chip designs require at least one re-spin. There, 75% of these designs contain functional or logical bugs. The increasing amount of embedded software implemented in integrated circuits further complicates verification. Studies, e.g. [Hum04], implicate that software still contains about 10 to 20 defects per 1,000 lines of code after compiling and testing is done. Remarkably, software companies have to spend nearly the same cost and time efforts on quality assurance like hardware manufacturers have to invest [Tas02].

Today, integrated circuits are mainly designed starting at the *Register Transfer Level* (RTL) using classical *Hardware Description Languages* (HDL) such as Verilog or VHDL. Studies have shown that up to 80% of the overall design costs are caused by verification when designing at RTL. The discrepancy between integrable design sizes and the actual verification productivity causes the so called "verification gap". This gap increases expo-nentially due to more and more complex and heterogeneous designs, which cannot be designed and verified efficiently at RTL anymore. Hence, today the Electronic Design Automation community is faced with an insufficient and inefficient approach for the design and the verification of SoCs.

One possible solution is a raising of the abstraction level towards the *Electronic System Level* (ESL). The ESL design paradigm has been developed by the academic and research community over the last years providing new design methodologies, proper tools, and special system design languages.

F. Rogin and R. Drechsler, *Debugging at the Electronic System Level*,
DOI 10.1007/978-90-481-9255-7_1,

Designing complex systems at ESL gets an increasing attention and application in industry, especially driven and supported by system providers such as ARM, or STMicroelectronics. At ESL the often textual specification of a SoC is coded into an executable specification in terms of a system model providing a virtual prototype. The system model enables an early system exploration and verification, and the parallel development of hardware and (embedded) software. Moreover, it acts as the common communication platform between system architects, embedded software developers, and hardware engineers along the value chain. Later, the system model is used as a *golden reference model* for upcoming design stages. *Transaction-Level Modeling* (TLM) is a promising modeling approach for ESL design. A TLM-based system model consists of several blocks that are connected by communication channels. The entire communication is described by function calls in terms of transactions. A transaction delivers arbitrary data between two blocks where only the needed functionality and the requested abstraction level (from untimed, pure functional descriptions up to cycle-accurate, mixed structural and behavioral descriptions) is implemented. A TLM description is accurate and fast enough to execute production software on it. Thus, it enables an early exploration and verification of the modeled system and its functionality.

Creating correct and reliable system models is an essential requirement for successful ESL design and all subsequent design steps. Although design concepts like object-orientation, borrowed from software development, and TLM boost the design productivity, the verification productivity falls behind. However, a simple reporting of a failure is not enough, anymore. In fact, detecting the failure-causing bug is an unsystematic and lengthy process that is increasingly becoming a bottleneck to chip design productivity. The challenge for successful ESL design is not only to find many errors during development early but also to provide methods to debug designed systems efficiently.

1 GENERAL OBJECTIVE OF THE BOOK

In this book, we develop a *systematic debugging approach* that enables a methodic search for errors while verifying ESL designs. Using this approach, errors can be detected and located efficiently at the different development stages of the system model. The proposed debugging techniques accompany each development stage starting as soon as first modules are available and continuing with subsystems until the entire system design is completed. In contrast to ESL verification flows, that predominantly focus on a late system level simulation, the presented approach allows to detect an error faster. Figure 1.1 shows the link between the proposed techniques and the different

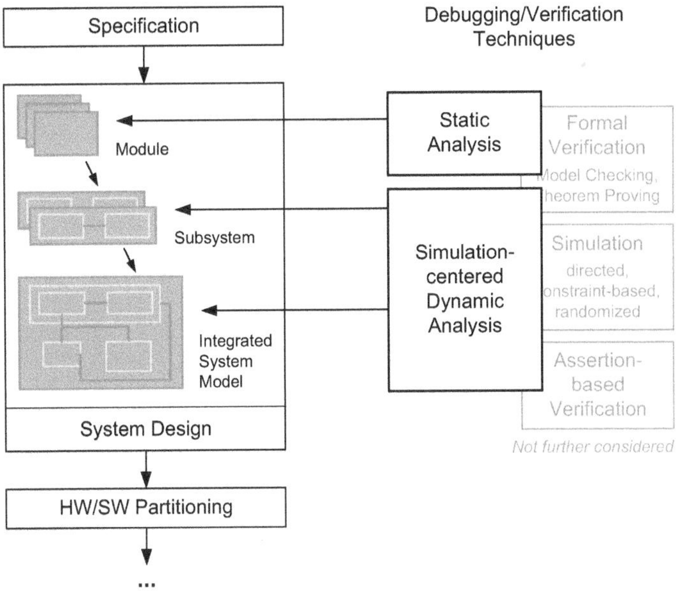

Figure 1.1: Debugging techniques in a simplified ESL design flow

development stages of the system model in a simplified ESL design flow. *Static analysis* starts as soon as first (non-simulatable) modules of the developed system are available. Removing an error in early development stages is cheaper and can be done easier. During the ongoing integration of subsystems to the system model, different *simulation-centered dynamic analysis techniques* facilitate debugging. They accelerate error search by improving controllability, observability, and evaluation of the simulation and the simulation results. So, the time between the insertion of an error and its correction is minimized, finally resulting in a reduced debugging time.

The systematic debugging approach supports the designer in a better utilization of the available debug capacity. To illustrate this fact, Figure 1.2 sketches the ratio between the bug-rate and the available debug capacity in a verification flow using only a late system level simulation. In contrast, Figure 1.3 depicts the improved utilization of the available debug capacity if static and dynamic analysis techniques are used during development, as proposed in this book.

The proposed debugging approach is applicable to any system level design language, such as *SystemC* [OSCI] or *SpecC* [GDPG01], and is independent of the level of abstraction on which the system model is described. That means that untimed as well as timed design descriptions could be used. However, some of the debugging techniques are better suited to a specific level

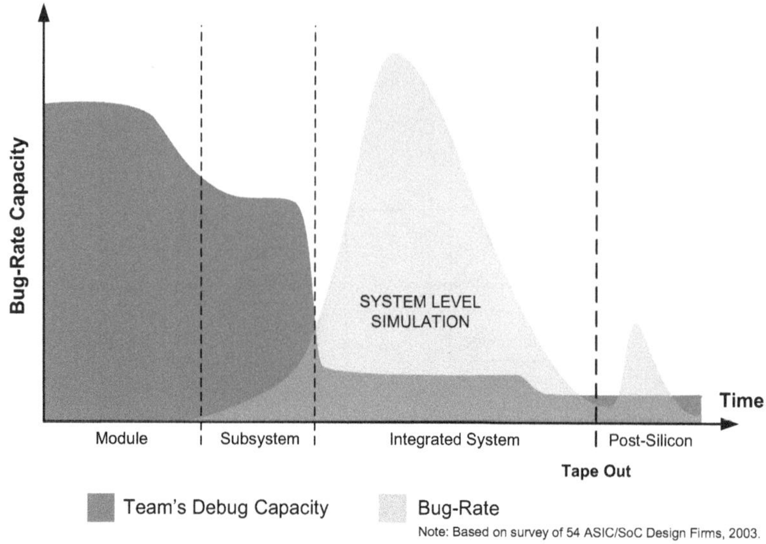

Figure 1.2: Verification efficiency using only late system level simulation (chart trend according to [Fos06])

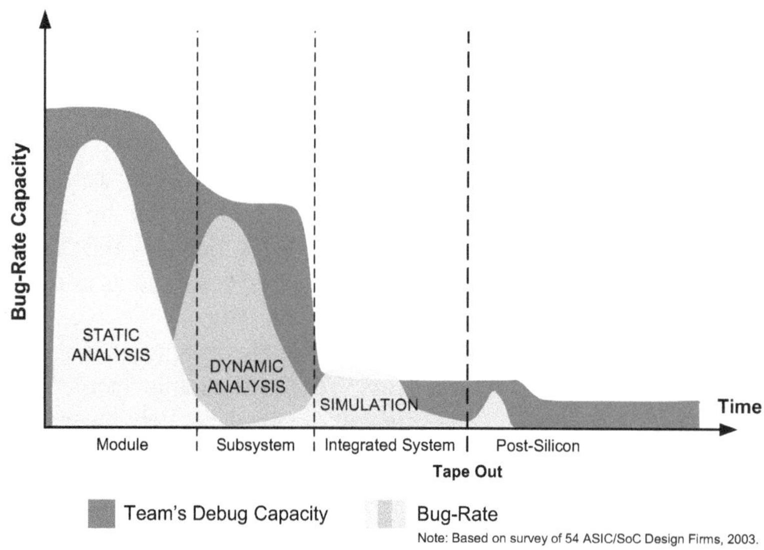

Figure 1.3: Improved verification efficiency using static and dynamic analysis techniques (chart trend according to [Fos06])

than others. A further focus has been placed on the industrial adaptability of the proposed techniques in terms of processible design sizes and complexities.

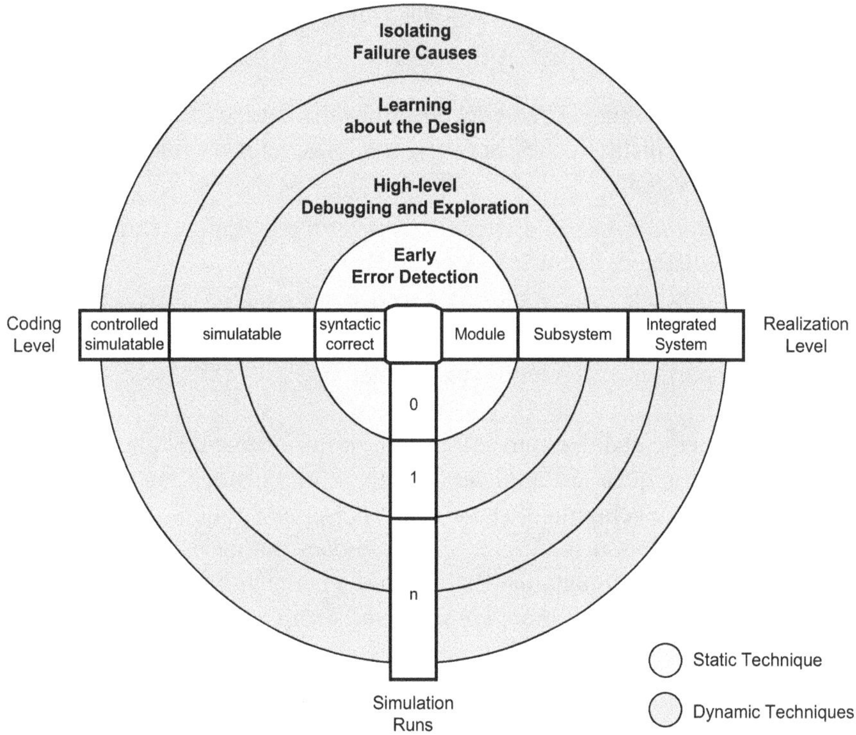

Figure 1.4: Hierarchy of proposed debugging techniques

2 SUMMARY OF CONTRIBUTIONS

Figure 1.4 correlates the introduced debugging techniques forming a hierarchy of co-ordinated reasoning techniques. The techniques accompany the development stages of a system model starting with deduction, i.e. *early error detection*, followed by observation, i.e. *high-level debugging and exploration*, next followed by induction, i.e. *learning about the design*, and closing with experimentation, i.e. *isolating failure causes*. The diagram in Figure 1.4 systemizes the utilization of each technique according to a set of defined preconditions for their application. These preconditions focus on the implementation progress while developing the system model. In contrast, the applicability of a particular debugging technique does not depend on the target architecture (e.g. processor, microcontroller), the communication design (e.g. bus, protocol, memory-mapped), the system level design language (e.g. SystemC, SpecC), or the abstraction level (e.g. untimed, timed). The implementation progress is judged by three classification criteria, i.e. the reached *realization level*, the number of needed *simulation runs*, and the achieved *coding level*:

- *Realization level.* This level describes the degree of model completeness that distinguishes between modules, subsystems, and the integrated system.
- *Simulation runs.* A different number of simulation runs allows to apply different debugging techniques.
- *Coding level.* The coding level defines whether a simulation and which type of simulation (uncontrolled vs. controlled) is needed to use a particular debugging technique.

Using the presented diagram in Figure 1.4, the system designer can determine which techniques are applicable at the current modeling state. At the beginning of the development, static analysis can be applied to first syntactical correct design modules. There, a module does not necessarily have to be executable and thus simulatable. Bit by bit, the realization level increases and the system model becomes more complex and complete. Modules are integrated to subsystems, and finally the model is completed at the actual abstraction level. Once, first components of the model can be simulated, dynamic analysis techniques enhance the analysis capabilities to locate and fix bugs that cannot be found statically, otherwise.

Example 1. *We assume the designer has coded an simulatable subsystem of the system model and wants to simulate (execute) it for the first time. According to Figure 1.4, two techniques, i.e. 'early error detection' and 'high-level debugging and exploration', are applicable.*

The proposed debugging techniques contribute to a systematic verification process of ESL designs. Each technique is prototypically realized in a SystemC-based design flow and is (mostly) evaluated in the field using industrial examples. Moreover, an example, that is used throughout the book, demonstrates the particularities and strengths of each technique:

- *Early error detection.* An analysis framework, called REGATTA, facilitates the flexible and fast generation of tools that statically analyze source code of arbitrary formal languages. The code quality is ensured by coding standards. Design flaws can be detected using more sophisticated analysis features, e.g. by using dataflow analysis techniques. Parts of this work were presented in [RF04], [RFSH05]. Based upon the REGATTA framework a static analyzer for SystemC, called SDAS, implements an industrial set of coding guidelines.

- *High-level debugging and exploration.* A debug flow guides the designer in debugging and exploration of system models at a higher abstraction level. Based upon open source tools and standards, an integrated debugging environment, called SHIELD, is proposed. It allows to debug arbitrary SystemC designs systematically. The papers [RFHO07], [RF+07], [RGDR08] are closely related to this chapter.
- *Learning about the design.* A new methodology automatically generates complex design properties for a simulatable system model using simulation traces. A prototypical tool, called DIANOSIS, deduces such properties. Complex properties improve design understanding, can be used as a starting point for formal verification, help to identify gaps in the test suite, and ease error detection. In [RK+08], [RK+09] parts of this chapter were published.
- *Isolating failure causes.* Using a series of controlled simulation runs allows to isolate the failure-inducing cause automatically in an erroneous system model. A prototype tool, called ISOC, adapts this automated debugging technique to SystemC, in order to narrow down failure causes in process schedules. A further application is the isolation of the minimal failure-causing difference in the simulation input. The paper [RDR09] is closely related to this chapter.

This book proposes an integrated approach that assists the designer in a systematic verification at the ESL. It combines and adapts existing and new debugging techniques. The techniques improve and accelerate error detection, observation, and isolation as well as design understanding. Hence, their continued application minimizes the number of errors that escape to the next development stage.

3 BOOK OUTLINE

This book is structured as follows:

In Chapter 2, the state-of-the-art of ESL modeling and verification is reviewed. Moreover, the system description language SystemC and a SystemC TLM2 example design, that is used throughout the book, are introduced.

The rest of the book discusses each of the debugging techniques in a separate chapter. Each chapter starts with the classification of the respective technique followed by a summary of the related work in the particular field.

The main part details the specific technique in general and their specific realization in a SystemC-based design flow. Each chapter closes with a demonstration of the particularities and strengths of the introduced techniques and is followed by an experimental evaluation using industrial designs.

Chapter 3 introduces static analysis as a deductive reasoning technique for error detection and code quality assurance of yet incomplete and non-simulatable system models. For this purpose, a static analysis framework is described which is subsequently extended to create a full static analyzer for SystemC.

For the simulation of first completed subsystems, Chapter 4 puts forward an observational reasoning technique which allows a systematic debugging and exploration of system models at a higher level. An integrated debugging environment is presented. This environment facilitates and accelerates error search by using debug patterns, a debugger being aware of the used *Specification and Description Language* (SDL), and a visualization component.

Multiple simulation runs form the basis for the introduction of an inductive reasoning technique in Chapter 5. Hence, different aspects and properties can be learned about the developed system model. Based upon a new methodology for automatic property generation, a practical implementation is shown. Property generation has been applied successfully to different industrial designs.

In Chapter 6 an experimental reasoning technique is proposed. This technique automatically isolates failure causes in system models using the delta debugging algorithm. It is applied to narrow down the failure-inducing difference in process schedules. Moreover, it is used to report the minimal difference in the simulation input that causes a SystemC simulation to fail.

Finally, Chapter 7 concludes and summarizes the contributions of this book.

Chapter 2

ESL Design and Verification

First, the state-of-the-art of modeling and verification at the ESL is briefly summarized in this chapter. As SystemC has been developed as a de-facto standard for system level design over the past few years, it is used for practical demonstration purposes in this book. Next, the introduced ESL methodologies and techniques are discussed under the SystemC perspective. Finally, our systematic debugging approach for ESL designs is described.

1 ESL DESIGN

The rising "design gap" demands for continuous improvements of the design productivity. One of the most critical issues is, how the available chip capacity, following Moore's Law, can be utilized by the chip designers. They have to develop even more complex integrated circuits and systems under fixed time-to-market and quality constraints. According to the ITRS report [ITRS07], a designer has a productivity of 125k gates per design-year using a state-of-the-art RTL-centered design and verification methodology. Here, designers describe an integrated circuit with an HDL (e.g. Verilog or VHDL) at RTL. There, the design is (almost) automatically synthesized down to circuit layout ready for fabrication. On the verification side, sophisticated verification tool suites are applied using techniques such as tracking and reporting information about the code coverage, or performing a constraint random simulation.

Now, we naively assume that a 10 million gates circuit shall be created from scratch. If the above mentioned design productivity of 125k gates is taken into account, 80 designers have to work for one year to complete the integrated circuit. Even more complex systems would require an unacceptable high number of designers or an improper long design time. Moreover, time-to-market, and rising cost and quality constraints demand for a new modeling and verification methodology.

Two basic approaches could help to improve the design productivity: The first one lifts the design level to a higher more abstract level above RTL. The second approach introduces a design reuse methodology by means of (third-party) IP components. Modern flows combine both approaches to maximize benefits.

F. Rogin and R. Drechsler, *Debugging at the Electronic System Level*,
DOI 10.1007/978-90-481-9255-7_2, © Springer Science+Business Media B.V. 2010

Designing above RTL is called ESL design. Generally, abstraction aims at the reduction of the effort to specify a desired functionality whereas unnecessary information is omitted. At the ESL designs are described by using higher level concepts such as communicating processes exchanging more abstract information. Here, the accurate timing or the parallelization of system functionality is not initially taken into account. In fact, the designer starts to specify the general function of the system and successively refines the specification until the concrete system, consisting of hardware and software, is fully implemented. Today, ESL design is supported by different SDLs where C/C++-based programming languages play a major role.

IP reuse is an important opportunity to improve the design productivity. The number of blocks to be defined from scratch becomes smaller if the designer can reuse existing IP components. IP blocks are taken either from former development projects or by using external IP providers. IP shall fulfill a number of requirements to enable a successful reuse. First, they should be of high-quality in terms of correctness and completeness. Second, the IP component interfaces have to be clearly defined and documented. Third, IP blocks should separate communication from behavioral parts to ease the integration into an existing design architecture.

1.1 ESL Design Flow

Currently, in the literature no universal design and verification flow for ESL design is documented. Rather, many different flow variants are stated. They are geared by customer and product requirements. The flow variants distinguish among each other by means of the introduced abstraction levels, use cases, and later application fields. The following section gives a short overview about important work that deal with the ESL design methodology. Baileys et al. [BMP07] give a comprehensive summary.

According to Gajski et al. [GV+94] the general ESL flow can be described by the three main steps: *specifying*, *exploring*, and *refining*. First of all, a functional specification of the desired system is developed. During the subsequent *exploration* phase, different design alternatives are compared in order to meet various requirements, e.g. performance, power consumption, or configurability. In the *refinement* phase more and more functional behavior is mapped onto a structural description consisting of hardware and software components. Subsequent refinement steps repeat exploration and refinement phases until the whole system is structurally specified. All other mentioned approaches adhere to this general procedure.

Kogel [Kog06] refers to use cases in terms of views to overcome the difficulties to specify an ESL flow at the level of abstractions. ESL modeling gears to many different domains, e.g. communication, time, structure, or behavior. Each domain can be described at particular levels of abstraction, e.g. the communication behavior can be modeled as transactions, bus functional models, or pin-accurate descriptions. Kogel concentrates on the purpose of the model by defining use cases. The *Functional View* (FV) use case is intended to create an executable specification of the application. The *Architects View* (AV) use case targets to the architectural exploration of the developed system to meet the desired requirements such as performance, power, or maintainability. The *Programmers View* (PV) use case provides a virtual prototype platform which could be applied for early embedded software development in parallel to the hardware design. Finally, the *Verification View* (VV) use case targets at cycle-accurate system modeling. So, an accurate performance evaluation prior to the design implementation or a TLM-RTL co-verification can be performed.

Teich and Haubelt [TH07] introduce an approach which summarizes important abstraction levels and views. These levels and views are passed during embedded system design. This approach distinguishes between *architecture* and *logic* abstraction levels in case of hardware designs. The *block* and *module* levels describe abstraction levels in the software development domain. Beside the classification of system models according to their levels of abstraction, two orthogonal views are defined. The *behavioral* view takes only the functionality of the system into account while the *structural* view details the communication between hardware and software components.

The SystemC TLM2 standard [TLM2] does not focus on abstraction levels. In fact, the standard mentions particular use cases, such as software development, software performance analysis, or hardware architecture analysis. These use cases are supported by two different coding styles, i.e. loosely-timed, and approximately-timed. Coding styles guide the designer in model writing using particular programming interfaces.

Fujita et al. [FGP07] document a design flow for system level designs based on C/C++-based system level languages. Starting with a textual specification an executable specification, mostly in terms of a program, is created. The resulting *functional system model* describes the general application and is mainly used to explore different functional design alternatives. During the *architecture design* phase, also known as *hardware/software partitioning*, the designer decides which functionality is implemented either into dedicated hardware components or software programs. Software plays an important role to allow for reusability, i.e. easy reuse of software components in terms of IP, maintainability, i.e. easy bug fixing and feature extension in the application field, and flexibility, i.e. easy replacement of software components. The exact communication between partitioned components, including processing order and parallelism, is detailed in the subsequent *communication design* phase resulting in the *communication model*. The software is usually implemented using a standard software development process. However, the functions that are assigned to hardware parts are further decomposed into structural units to take the target hardware platform into account. Finally, within the *implementation design* phase, the hardware parts of the communication model are synthesized into an RTL design using *high-level ESL synthesis techniques*. Several tools automate this step. Nevertheless, in practice it is usually still a manual or at most a semiautomatic step. Finally, the RTL design is implemented which comprises well automated steps such as logic or layout synthesis. The developed software descriptions can be directly extracted from the communication model and are compiled into executable machine code.

Bailey et al. [BMP07] published one of the first books that summarizes state-of-the-art in ESL design and verification methodology. One important contribution is the definition of a taxonomy for the ESL design space and the definition of common terms used in this domain. The proposed ESL taxonomy contains the five axes concurrency, communication, configurability, temporal, and data. This taxonomy spans a space for the definition and classification of ESL specification languages, design flows, and tools. Bailey et al. divides the ESL design flow into six main development steps: *specification and modeling*, *pre-partitioning analysis*, *partitioning*, *post-partitioning analysis and debug*, *post-partitioning verification*, and *HW/SW implementation*. In the specification and modeling step the designer translates the initial informal specification document successively into various executable models. During the pre-partitioning analysis step, the algorithmic design space is explored taking different constraints into account, e.g. time, space, power, complexity, or time-to-market. The partitioning process defines which parts of the functionality are implemented either in software or hardware. The effect of hardware/software partitioning is explored in the post-partitioning analysis and debug step. If necessary, a re-partitioning takes place to better meet the requirements. Then, the following post-partitioning verification step ensures that the originally intended behavior of partitioned software and hardware components is preserved. Finally, the HW/SW implementation step creates synthesizable RTL models as well as the productive software that shall run on the target hardware.

Summarized, all aforementioned work can be mapped more or less to the idealized ESL design flow sketched in Figure 2.1. Rather than this idealized and simplified linear top-down flow, a real flow is a mixture of bottom-up and top-down procedures. Moreover, the development of an integrated system consists of many refinement steps and backtracking paths that shall be not further detailed here. In reality, parts of the model are usually implemented at different abstraction levels at the same time. So, some parts could be specified

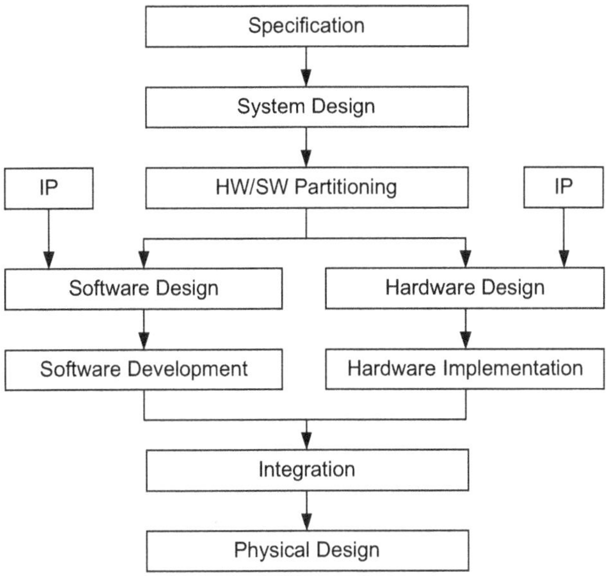

Figure 2.1: Idealized ESL design flow (taken from [BMP07])

at a more structural level while other parts are initially described at the behavioral level. The model can also contain loosely-timed specified components as well as cycle-accurate blocks in parallel. Nevertheless, a designed system always passes different levels of abstraction starting from the most abstract specification. Then, this specification is successively refined to the lowest most accurate level.

Figure 2.2 illustrates the design levels while implementing the hardware parts of a SoC design. It shows an idealized top-down procedure following a SystemC TLM oriented design methodology:

Starting with a textual specification, the *algorithmic design* is specified. Thus, the basic functionality of the developed system is explored while the distinction between hardware and software components as well as the timing is not yet taken into account.

Based upon the algorithmic specification, the design is refined to a TLM-based system model. The model consists of different blocks that are connected by communicating channels. Channels are used to exchange data between these blocks in terms of *transactions*. During development, the system is normally refined using different timing levels. At the *loosely-timed* level, the system model consists of functionally accurate components describing the system platform. Synchronization between components is only supported at a coarse-grained synchronization level specifying the general scheduling procedure, e.g. the correct order between produced and consumed data.

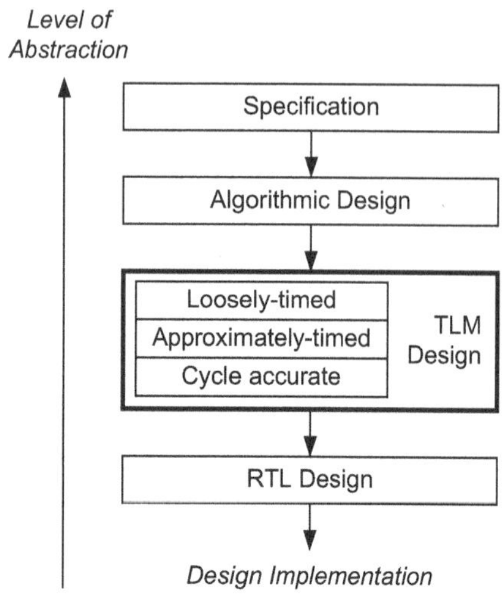

Figure 2.2: Designing the hardware part of an SoC

Moreover, communication channels are used for synchronization purposes, e.g. to handle interrupt signals. The loosely-timed abstraction level is the basis for early embedded software development and debugging where unnecessary implementation details are left out. The aforementioned PV use case relates to models at this level. At the next level, i.e. *approximately-timed*, the timing is refined which enables a preliminary performance estimation and a coarse power analysis. A model at this level is available earlier than the RTL model and can facilitate reasonable design decisions for the later RTL design. This abstraction level corresponds to the *PV plus Timing* use case. A *cycle-accurate* model describes the design behavior at the level of clock cycles. This abstraction level is used e.g. for final HW/SW partitioning, HW/SW co-verification, or a TLM-RTL co-simulation.

Transferring a system model into an RTL design is accompanied by a change of the design language. While system models are described using an SDL such as SystemC or SpecC, RTL designs are usually implemented using HDLs such as VHDL or Verilog. An RTL design contains all details of the hardware components and can be automatically and efficiently synthesized. Hardware synthesis comprises the design implementation, such as logic or layout synthesis, and is well supported by state-of-the-art tools and design flows. Detailed information to hardware design and synthesis can be found for instance in [Ash06], [Pal03].

1.2 System Level Language SystemC

C/C++-based languages have been playing a major role in embedded software development for a long time. So, it was apparent to use the same or some similar language to specify the hardware parts of an ESL design and to write test benches, as well. However, C/C++ do not provide constructs to deal with concurrency, communication, or timing necessary for designing hardware. Several extensions were proposed over the past few years to make C/C++ ready to be used in ESL design. Today, two C/C++-based SDLs are mainly used: SystemC and SpecC. Both languages provide similar language constructs for hardware design. The major difference is the underlying language. While SpecC bases upon C, SystemC is a C++ class library. Hence, SystemC has an object-oriented nature which facilitates the implementation of abstract, modular, and reusable system models. SpecC as well as SystemC are freely available which ease their application and distribution. Additionally, IEEE approved SystemC as a standard in 2005 [IE+06] which has increased the acceptance of this language in industry. The standardization process has been significantly driven by the OSCI. In this book, SystemC is used for demonstration purposes. However, the proposed debugging technique are applicable to any other system level language.

SystemC extends C++ by notions of modules, concurrent processes, simulation time, and communication mechanisms. These features support ESL design while comprising the full power of C++. A SystemC design is created by a hierarchy of (nested) modules which are instances of the class `sc_module`. Modules communicate through ports, interfaces, and channels. An interface provides several methods to access a channel whereas the methods are called through a port. A channel separates communication and computation. SystemC provides a number of predefined channel types ranging from simple wires to more complex communication mechanisms like FIFOs. Furthermore, designers can derive their own channel types. The main elements of computation are represented by concurrent processes that are either method (`SC_METHOD`) or thread processes (`SC_THREAD`). A process has to be specified inside a module. A thread process terminates never and suspends its computation by calling a `wait` statement. In contrast, a method process completely executes in zero time, every time it is triggered.

An integral part of the SystemC library is the event-driven simulation kernel which implements a cooperative multitasking approach and divides in two main phases:

- *Elaboration.* Execution starts at the `sc_main` function. First, all modules are instantiated and bound together. After this phase a further change of the model structure is not possible. Then, the call of `sc_start` launches the simulation.

- *Simulation.* During simulation the SystemC kernel executes each runnable process one by one in a non-preemptive fashion. A process suspends again either when it was completely processed (method process) or the process calls a time consuming statement (thread process). There, the SystemC simulation kernel works like an HDL simulator using the notion of delta cycles. A delta cycle is orthogonal to the simulation time where the time is only advanced, when no more processes are runnable at the current time step, i.e. at the current delta cycle. The use of delta cycles emulates the parallel execution of processes. A process, running at the current delta cycle, can wake up other processes through an immediate event. Another variant is a signal update which will be executed during the next delta cycle. Again, these processes can trigger further ones at the subsequent delta cycle.

A detailed introduction into SystemC is given in the SystemC Language Reference Manual [IE+06] or can be found in [BDBK08].

1.3 Transaction Level Modeling in SystemC

With the publication of the OSCI TLM2 standard in 2008 [TLM2], SystemC took a major step forward to facilitate early system exploration, co-design of hardware and software, and platform-based design and verification. Moreover, a TLM modeling style improves the reusability of IP components between different IP providers. A standardized *Application Programming Interface* (API) allows the separation of computation inside components from the communication among components. The communication is modeled by channels where a transaction is released by calling interface methods of the channel. Unwanted communication details are hidden in the channels that are iteratively refined during modeling. Hence, system design starts with an abstract TLM description which enables a faster simulation as a comparable but more detailed RTL model. A predefined TLM interface is implemented by a high-level communication model initially describing only the exchange of messages or data between components. During development, the same interface can be implemented by a cycle-accurate *Bus Functional Model* (BFM) specifying the behavior of each pin later implemented in hardware. A BFM enables the designer to write test cases in terms of abstract method calls at TLM level. Then, the BFM translates the method call into particular input stimuli at the RTL site.

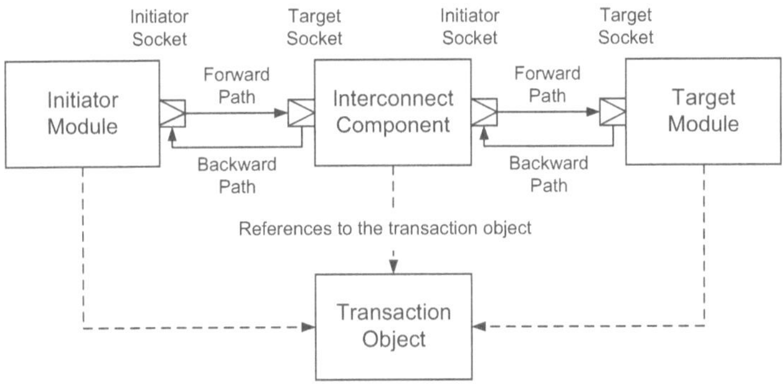

Figure 2.3: TLM notion in an example (taken from [TLM2])

The SystemC TLM2 standard distinguishes between coding styles and interfaces instead of defining abstraction levels for each particular use case. A use case can be for instance software development, software performance analysis, or hardware verification. The coding styles guide the designer in system modeling where the interfaces define low-level programming mechanisms. The TLM2 standard defines two coding styles: *loosely-timed*, and *approximately-timed* that are supported by particular blocking and non-blocking transport interfaces. Figure 2.3 documents an exemplary TLM design. A module can act in three different ways:

- *Initiator*. This module type creates new transactions and passes them to the channel by calling a predefined interface method.
- *Target*. A module of this type receives transactions and executes them according to the target module task.
- *Interconnect*. This component type forwards a transaction and possibly modifies it. So it acts as initiator and target at the same time.

The transportation path of a transaction that is going from an initiator to a target, is also called the *forward path*. The opposite direction is named the *backward path*. Over the backward path, the target informs the initiator about the transportation state. Either the modified transaction object is returned or a specific backward method is called explicitly. Two socket types encapsulate the connection between components. The *initiator socket* enables interface calls on the forward path by a *port* and on the backward path by an *export*. The *target socket* offers the same mechanism in case of a backward path. For a complete documentation of the TLM2 standard refer to [TLM2]. A comprehensive overview about the state-of-the-art of TLM design in SystemC can be also found in [Ghe06].

2 ESL VERIFICATION

Verification of functional design correctness has become the dominating cost and time factor in electronic system design, today. The correctness of a system has to be checked at each design step to identify errors as early as possible during system development. The later an error is detected, the more the costs for its correction increase. To cope with the rising verification complexity, a wide range of advanced techniques is used such as static analysis, debugging techniques, formal verification, assertion-based verification, constraint-random simulation, automatic test pattern generation, or model-based specification approaches. As already discussed on page 9, each abstraction level provides a defined degree of detail to fulfill a certain modeling or verification task, e.g. to measure the software performance or to explore different architectural choices. Starting with a system model, the model is refined step-by-step until the system is completed in the final software and hardware. An efficient abstraction mechanism for ESL design is provided by the introduction of SDLs together with a TLM design methodology. TLM facilitates the separation of computation and communication and allows to refine different design parts independently of each other (see Section 1.3 on page 16).

It is promising to extensively check higher level, less complex, descriptions before refining them. So, the detection and correction of an error is more efficient. A system design passes many refinement steps where in case of an error the design has to be fixed. Sometimes taken refinements are backtracked starting at a previous step, again. *Equivalence checking* ensures the equivalence between the current implementation and the specification. Does the implementation satisfy the specification, it will become the specification for the following design step. Writing the initial specification is a time-consuming and error-prone process. An imperfect specification increases the probability for *false positives*, i.e. pointing out non-bugs as bugs, or *false negatives*, i.e. failing to find real bugs. So, it is crucial to start with a correct and preferably complete specification which describes all important and desired aspects of the system.

The availability of an ESL design flow does not only allow the modeling of complex hardware. Moreover, the system model enables the designer to develop and to verify embedded software in parallel with the hardware. There, the hardware blocks can be exercised from the software programmers perspective. So, many functional errors can be found before the actual hardware is available. A system model has a further advantage. During development, parts of the system are replaced by their particular implementation (in hardware) whereas the rest becomes the verification environment. Thus, the hardware designer can be relieved from any additional writing of the verification environment. Such a verification approach is applicable to find system-wide problems instead of verifying the correct function of a particular hardware block. It is also known as golden reference model verification where the system model defines the specification for the hardware components.

System correctness can be ensured by two different approaches: dynamic techniques, i.e. *simulation*, and static techniques, i.e. *formal verification*. In between, there is a third approach, that combines simulation and formal verification using *assertions*. It is called *semi-formal verification*. Moreover, there are further verification techniques, such as prototyping or emulation, but these techniques shall not be discussed in this book.

2.1 Simulation

Simulation is still the predominant verification technique in system and hardware design. It is easy to use, scales very well with arbitrary complex designs, and detects many functional errors very efficiently. As soon as an executable specification is available, it can be simulated easily. Besides the check of functional correctness, simulation is used to identify performance problems, inconsistencies, or specification holes. Apart from these advantages, the complexity of current designs prevents an exhaustive simulation to prove system correctness in a limited time [ARL00], [ITRS07]. Here, an increasing number of state variables let exponentially rise the number of input patterns, and thus the number of required simulation runs. So, the quality of simulation-based approaches highly depends on the quality of available input patterns to gain a satisfying coverage of the design functionality. Due to its incompleteness simulation is also named *validation*.

2.2 Formal Verification

To overcome the limitations of simulation, formal verification techniques were developed [Kro99]. Formal verification proves the correctness of a system using a mathematical model of the system behavior. This technique is exhaustive, and thus ensures correctness. Three different approaches are used in formal verification: *theorem proving* [WOLB92], *equivalence checking* [KPKG02], and *model checking*, also called *property checking* [CGP00].

In theorem proving the verified system (the implementation) and the specification are described in a higher-order logic. With the help of axioms and inference rules, a theorem prover supports the designer in reasoning that the implementation implies the specification. Since higher-order logics are generally undecidable, a fully automated proving is not possible. The required manual intervention has prevented a widespread application of theorem proving in the industrial field, so far.

Equivalence checking verifies the equivalence between implementation and specification. Normally, this technique verifies the correctness of (automatic) synthesis steps, e.g. by proving the equivalence between an RTL design and the synthesized gate level description. A common approach uses a miter circuit where two automata representations, one for the implementation and one for the specification, are combined. Then, the outputs are checked for equivalence while feeding in the same input data [Bra83]. The equivalence checking problem can be solved by transforming the miter circuit into a SAT instance. Today, equivalence checking enjoys a wide distribution in semiconductor industry especially due to its good automation and easy handling.

Model checking bases upon a finite state space model of the verified system, e.g. a labeled transition system or a finite state machine. The specification is formulated in terms of a set of properties in some temporal logic. Then, a model checker formally proves the validity of each (possibly unbounded) property on the model. *Bounded Model Checking* (BMC) limits the number of time frames a property is checked. Since a system usually responds after a finite time limit, BMC is quite efficient. It is successfully applied in the semiconductor industry, e.g. [BB+07]. A restriction of model checking is the limitation of a straightforward model translation (design abstraction) for large designs due to the so called *state space explosion* problem. Assuming that an example design consists of many thousand lines of code and for instance 5,000 32 bit integer variables. In that case, a Boolean reasoning technique has to handle 5,000 * 32 = 160,000 Boolean signals over many thousand program states. This is a potentially infeasible number to be currently handled by formal verification tools. Hence, classical model checking cannot verify arbitrary complex systems. New approaches try to use sophisticated abstraction techniques to handle the state explosion problem. *Satisfiability Modulo Theories* (SMT) solvers [NOT06] process word-level information directly. Another problem of model checking results from verifying only an abstracted model of the design. Hence, an erroneous abstraction process can produce false negatives as well as false positives. Moreover, a wrong specification could show that the system works as specified, but not as primarily intended. Finally, there are problematic structures, such as multipliers, that cannot be formally verified easily.

2.3 Semi-Formal Verification

Semi-formal verification combines simulation (see Section 2.1 on page 19) and formal verification (see Section 2.2 on page 19) and is also known as *Assertion-Based Verification* (ABV) [FKL03]. Assertions are temporal properties that concisely capture design intent. They are dynamically monitored during simulation. Two languages gain importance for applying ABV techniques and allow the exact description of temporal-logic expressions: *SystemVerilog Assertions* (SVA) [IE+05a] and the *Property Specification Language* (PSL) [IE+05b]. To enable a fast and easy creation of an ABV test suite, the *Open Verification Library* (OVL) [OVL] captures typical design behavior in terms of assertion checkers.

ABV does not prove the specified property like in formal verification. For this reason, ABV can be applied to arbitrary complex designs. Additionally, assertions improve design observability, and thus facilitate the debugging process. In case of an error, an assertion directly points to the location the problem was recognized. This approach is also called white-box verification instead of the simulation oriented black-box approach. In simulation the response of a system is checked for correctness only at the output interfaces.

2.4 Verifying SystemC Models

SystemC only supports simulation natively while formal and semi-formal verification approaches are subject to current research. The following section summarizes important work in the particular fields.

2.4.1 Simulation in SystemC

The simulation-based validation is inherently supported by SystemC. Every standard C++ compiler, e.g. the GNU C++ compiler, is able to compile an arbitrary SystemC description into an executable program. Such a program can be directly simulated since the simulation kernel is an integral part of the SystemC library.

The processing of a SystemC simulation is particularly influenced by the style of coding. Several coding styles allow a different modeling of communication and synchronization between concurrent processes. SystemC FIFOs and semaphores introduce communication and synchronization points into the application. This style is formalized by *Communicating Sequential Processes* and *Kahn Process Networks*. It permits a completely *untimed* modeling style without the need for advancing the time during simulation. A coding style where each process yields control at a certain point in time is called *timed*. There, different timing levels can be distinguished such as loosely-timed, approximately timed, or cycle-accurate. A timed coding style is supported in SystemC by explicit synchronization statements.

Simulation uses directed or randomized tests to compare the expected results with the observed system behavior. A directed test validates only a very certain functionality by using an explicitly specified stimulus pattern. Usually, it is manually written by the designer which is a time-consuming and erroneous task. A major improvement in test bench automation was achieved by the introduction of constraint-based random simulation [YPA06]. Based on a set of user-defined constraints, stimulus patterns are generated automatically. Hence, in a short time many more scenarios can be created and tested if compared to directed tests especially in case of corner cases. Constraint random simulation is supported in SystemC by the *SystemC Verification Library* (SCV) [SCV]. Moreover, this library provides other sophisticated verification features such as transaction recording and monitoring, or data introspection capabilities for arbitrary data types. Some work extents the SCV library. Große et al. [GED07] add bit operators and improve the integrated constraint solver to guarantee a uniform distribution of constraint solutions. Another work [GWSD08] introduces an approach that resolves contradictions between constraints automatically.

Based upon the SCV, various work propose verification frameworks or methodologies which allow for an easy, fast, and reusable simulation-centered functional verification of SystemC designs, e.g. [SMA04], [PC05]. These frameworks provide a unique verification infrastructure. They facilitate test bench creation and adaptation to a new design under verification, supply sophisticated coverage mechanisms, or integrate a constraint-random stimulus generator.

Functional code coverage techniques allow the designer to control simulation effort. Here, metrics provide a basis for decision in order to determine whether a SystemC design was simulated sufficiently. Various metrics measure code coverage using branch, statement, or condition coverage for instance. Traditional software metric tools for C++, such as gcov [GCOV], can be also applied to SystemC. The drawbacks of these tools are amongst others that coverage analysis also includes the SystemC kernel library routines. Furthermore, a C++ coverage tool does not know the SystemC semantics which requires a manual and tedious extraction of interesting coverage data. Große et al. [GPKD08] propose an approach that measures the quality of SystemC test benches in terms of a control flow metric. Therefore, an instrumented SystemC description collects coverage information during simulation. Subsequently, coverage information are analyzed whether the defined coverage goal has been already reached.

2.4.2 Semi-Formal Verification in SystemC

Since the SystemC standard does not define a native support for ABV, various approaches propose an integration of assertions into SystemC.

One of the first approaches was introduced by Ruf et al. [RHKR01]. It uses finite linear temporal logic to specify properties. These properties are translated into a special kind of finite-state machine, i.e. a monitor, in a preprocessing phase. During simulation a monitor dynamically checks simulation behavior and reports each violation immediately.

The integration of SVA into SystemC is proposed by Habibi and Tahar [HT04]. An SVA expression is translated into an external SystemC module. This module acts as a monitor to all signals involved in the formulated assertion. Using the symbol table of a compiled SystemC design allows to connect the external monitor component with the design automatically. A so called design updater modifies the SystemC code in order to link design code and assertion monitor. During simulation, all SVA expressions are checked on-the-fly. A subsequent work [HT06] presents a model-driven design approach. Here, properties are modeled by extending UML sequence diagrams in order to check transaction properties of a SystemC TLM design. Therefore, the UML model of a PSL property is translated to an *Abstract State Machine* (ASM) description. Then, the ASM description is checked against the SystemC model which has to be available as an ASM model, as well. Both models are further translated to SystemC code where the property is compiled into a C# monitor.

Große and Drechsler [GD04a] present a method where a property (assertion) is translated into a synthesizable SystemC checker. The checker is embedded into the original SystemC description. Due to the synthesizable characteristics of checkers, they can be also evaluated after fabrication of the system.

Bombieri et al. [BFF05] evaluate the adoption of PSL to a SystemC-based TLM verification flow. They distinguish between two techniques, "properties re-use" and "properties refinement". Properties re-use is utilized to ensure the functional equivalence of two different SoC descriptions. Properties refinement denotes the process of translating a PSL property, written for a more abstract design block, to the refined block. There, the PSL to SystemC checker translation is based upon the approach described in [DG+05].

Niemann and Haubelt [NH06] allow the concise formulation of assertions for TLM designs by using SVA expressions and interpret transactions as Boolean signals. To prevent a modification of the original source code, an aspect-oriented programming technique is used. Hence, transaction recording statements are weaved into the original SystemC design applying the AspectC++ compiler presented in [SLU05]. Then, the instrumented code is compiled and simulated. Transaction activity is written into a *Value Change Dump* (VCD) trace file. Subsequently, the VCD file is translated into a Verilog module which is simulated together with the formulated set of SVA assertions in a classical HDL simulator.

Kasuya and Tesfaye [KT07] introduce a native assertion mechanism for SystemC called NSCa. The designer creates assertions by using cycle-level temporal primitives comparable with assertions written in SVA or PSL. NSCa assertions are written either as a separate file or they are embedded into the SystemC design. Moreover, the provided temporal primitives are applicable for higher levels of design abstraction, i.e. at the algorithmic or the TLM level using events and a simple queue model.

The largest problem for a widespread application of ABV techniques in SystemC is the missing standardization. The application of the introduced approaches

- is limited to a subset of SystemC, e.g. synthesizable descriptions [GD04a],
- introduces a new proprietary assertion language [KT07],
- integrates assertions only indirectly using additional tools [NH06], [RHKR01], [BFF05], [HT04], or
- does not completely support the underlying assertion language [HT06].

2.4.3 Formal Verification in SystemC

Formal verification gets a rising interest in SystemC design. Apart from the object-oriented nature of SystemC and its event-driven simulation semantics, the integration of software and hardware modules in the same system model makes the formal verification of SystemC designs a challenging task. Various approaches provide a formal simulation semantic for SystemC, e.g. [MR+01], [HT05], [Sal03]. However, a full formal semantics does not yet exist which complicates the development of formal techniques.

In [DG02] Drechsler and Große propose a reachability algorithm for sequential circuits described in SystemC. The algorithm bases upon *Binary Decision Diagrams* (BDD). Due to the limitations to synchronous sequential circuits, more abstract system descriptions are not supported by this approach. Based on the presented symbolic reachability algorithm, an approach for the formal verification of properties specified in *Linear Temporal Logic* (LTL) is given in [GD03].

Kröning and Sharygina [KS05] introduce a technique that automatically partitions a system model into a synchronous hardware part and an asynchronous software part. Partitioning is performed by a syntactical distinction of thread types (combinational vs. clocked threads). Labeled Kripke structures formalize the semantics of SystemC. Due to the partitioning step, the verified model contains fewer transitions, and thus is more efficiently verifiable.

Habibi and Tahar translate SystemC models into an intermediate representation using *Abstract State Machines Language* (AsmL) in order to support model checking [HT06]. Using AsmL, the complexity of a SystemC design can be radically reduced which enables the verification of complex designs such as a PCI bus [OHT04]. The correct mapping between SystemC and AsmL is proven in [HT05].

Moy et al. enable the verification of SoCs described at transactional level. In [MMM05a] they introduce their toolbox LusSy that translates a SystemC design into a set of parallel automata that rely on a proprietary intermediate representation called HPIOM. HPIOM defines an executable formal semantics for TLM-based SystemC descriptions. Furthermore, it is connected to various formal tools such as the symbolic model checker LESAR. The translation rules are integrated into the GNU C++ compiler front end.

Herber et al. [HFG08] present an approach that defines the semantics of SystemC by a mapping from SystemC descriptions into Uppaal timed automata. These automata enables model checking in order to verify properties such as liveness, deadlock freedom or compliance with timing constraints.

The weakness of all formal verification approaches is their limitation in handling arbitrary complex SystemC designs. A missing formal semantics for the full language restricts the application of formal techniques in an industrial context. Mostly, the presented approaches support only a subset of SystemC features. Especially the use of object-oriented software concepts and the crucial pointer semantic cause the same problems model checking of C++ programs has. Although formal verification approaches promise to prove the correctness of system designs, particularly the complexity combined with a rising ratio of software prevent their widespread and straightforward usage. So, classical simulation still remains the means of choice. Nevertheless, new techniques, such as ABV, bridge the gap between simulation and formal verification, and try to combine the benefits of both worlds.

3 OUR DEBUGGING APPROACH

At first, this section defines common terms used in the context of debugging in this book. Next, the general debugging process is sketched. Afterwards, our debugging approach, consisting of various debugging techniques, is described. Finally, an example which is used throughout the book is introduced.

3.1 Terms

Debugging is the process of detecting and fixing a defect that has caused an observed failure. To clarify the different terms used in this context, i.e.

defect and *failure*, we want to define them in the same way as Zeller has done in [Zel05]:

- A *defect*, also known as *bug* or *fault*, is a faulty location in the program code that can cause an infection.
- An *infection* is an error in the program state that can cause a failure if this state is propagated.
- A *failure* is an externally visible situation in the program behavior that differs from the expected behavior.

It should be noted that a failure can be produced by a combination of defects where only the combination leads to an infection. Simultaneously, the same defect can produce many different failures.

3.2 General Debug Process

Debugging in the hardware as well as in the software domain can be mapped to the following general procedure:

1. Exactly describe the observed problem. Sometime, you can already catch the problem cause while revising the erroneous situation in detail.
2. Ask the question whether the problem could be a software failure?

   a. If yes

      reproduce the failure by simplifying and automating the failure-causing test case,

      locate the failure-causing bug, e.g. constitute a hypothesis and test it, observe the program state, isolate the correlation between cause and effect, and fix the bug.

   b. If no,

      Find the problem cause and fix it. The defect could be for instance a hardware problem, a bug in the used compiler, or a simple misunderstanding in handling the software.

3. Verify the effectiveness of the bug fix.

   a. Could the entire problem be solved?

   b. Are there new, formerly unknown problems?

3.3 Hierarchy of Debugging Techniques

The main objective of the book is the development of a *systematic debugging approach* that enables a methodic search for errors while verifying ESL designs. So, errors can be detected and located efficiently at the different development stages of the system model. As a result, the approach contributes to an improvement of the overall verification productivity. It starts with first testable modules. Here, static analysis identifies coding flaws and potential failure causes. At the end, a controlled dynamic analysis automates debugging of the completed system model. The application of one of the debugging techniques does not depend on the current used abstraction level (see Figure 1.4). So, static analysis can be utilized on untimed as well as cycle-accurate model descriptions. Rather, the classification of the techniques is made by the number of needed simulation runs, the reached realization level, and the achieved coding level (see Figure 1.4). The debugging techniques form a reasoning hierarchy as proposed by Zeller [Zel05]. Zeller introduces a hierarchy of program analysis techniques to provide a classification of debugging techniques especially used for software programs. This hierarchy shall help the programmer to systemize the overall debugging process:

- *Deduction.* Deduction denotes the reasoning from the statically analyzed program code to the concrete program run. Generally, it means the reasoning from the general to the particular. This reasoning technique bases upon an abstraction of the program that is used to deduce particular properties.
- *Observation.* Observation is the first dynamic analysis technique. There, a particular program run is used to explore arbitrary program aspects. In contrast to deduction, this technique observes the results of a single program run and tries to find approximations based on that run.
- *Induction.* In general, induction tries to infer from particular observations to the general program behavior. Actually this means, that many program runs are merged to a general abstraction which holds for all runs.
- *Experimentation.* Experiments are used to search for the cause of an observed failure systematically. A series of experiments is created that tests a hypothesis with the goal to reject or confirm it. As a result, a precise diagnosis shall be isolated using a number of controlled program runs.

In this book, the reasoning hierarchy shall be adapted to the specific requirements and conditions of debugging ESL designs in terms of SystemC model descriptions:

- *Deduction*. As soon as the designer has written syntactical correct source code, hypotheses can be deduced from the code without the need for simulating the model. A static analyzer for SystemC, which is based on a generic static analysis framework, performs different analyses, This concerns an analysis of coding standards and the check for functional errors.
- *Observation*. The observation of a particular simulation run of the developed system model is supported by a high-level debug flow. Various sophisticated debugging and exploration features support the designer in exploring the actual simulation state. A SystemC debugger, debug patterns, and visualization features allow a fast error detection, location, and correction.
- *Induction*. Multiple simulation runs generate a set of simulation traces. The traces are used to deduce from a number of concrete runs general abstractions by means of design properties. Properties, that are generated on the golden reference model, can be later used to check the functional correctness of more refined design models, e.g. the hardware part at RTL.
- *Experimentation*. Using a set of controlled simulation runs allows to isolate failure-inducing causes in a system model. The delta debugging algorithm of Zeller and Hildebrandt [ZH02] is adapted to SystemC to automate debugging of design descriptions.

Figure 2.4 illustrates the classification of the introduced debugging techniques in an ESL design flow (see Figure 2.1). There, a technique shall be used as soon as the preconditions for its usage are given (see Figure 1.4). Each technique is implemented by a prototypical tool for SystemC to demonstrate its effectiveness and efficiency. Generally, the techniques aim at a fast and easy location of errors in the system model.

3.4 SIMD Data Transfer Example

An example, that is used throughout the book, shall demonstrate the particularities and strengths of each introduced debugging technique. The example models the simplified data transfer between the cache of a *Processor Unit* and a global *Shared Memory* of an *Single Instruction Multiple Data* (SIMD) processor design. Figure 2.5 shows the general architecture of the

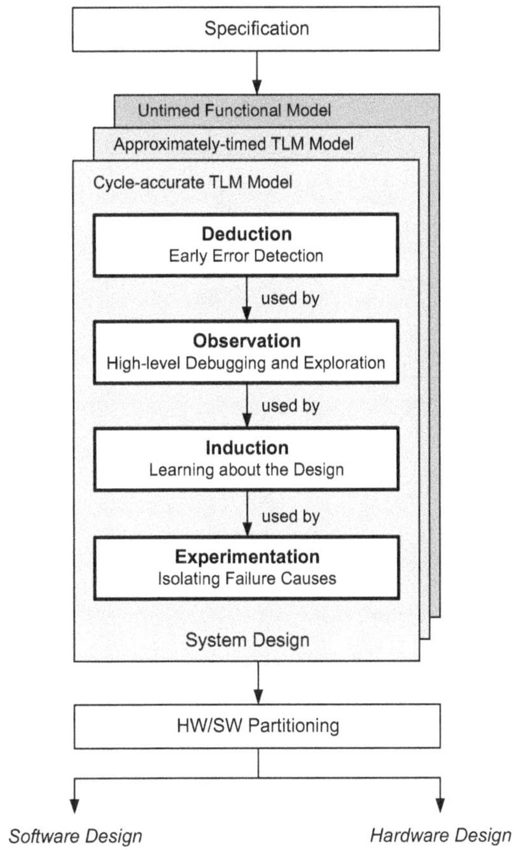

Figure 2.4: Reasoning hierarchy in a simplified ESL design flow

SIMD data transfer example. Here, only a single processor core of the entire processor design is considered. The synchronization mechanisms between the different cores are omitted for simplification reasons. The data transfer is represented by generic payload transactions conforming to the SystemC TLM2 standard [TLM2]. The memory access is handled by a *Load/Store Controller* attached to the particular processor unit. Here, an integrated arbiter controls and handles the read and write transfers from/to the shared memory. Data are either read or written, i.e. there is a mutual exclusion between read and write accesses due to a specified single port cache structure. An internal buffer in the load/store controller decouples memory and cache. The external *Monitor* component supports debugging tasks. It records the transaction activities and writes them into a VCD dump file.

Figure 2.6 depicts an example read/write data transfer between the processor cache and the shared memory. Each data transfer starts with a read

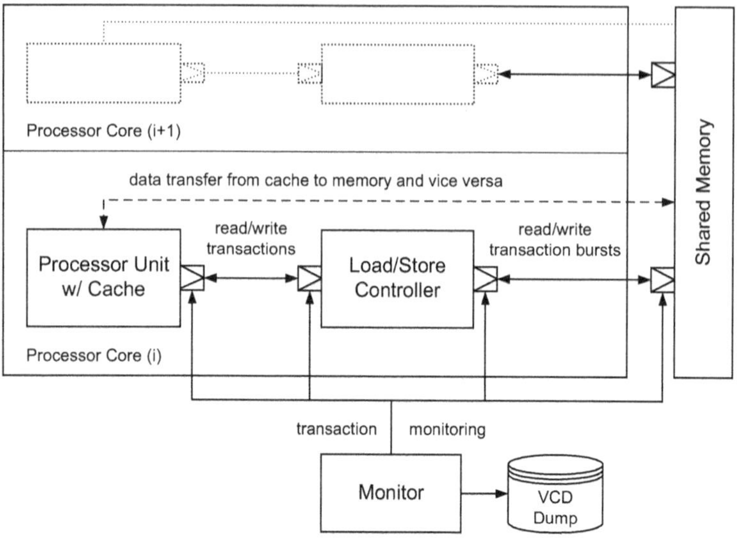

Figure 2.5: General architecture of the SIMD data transfer example

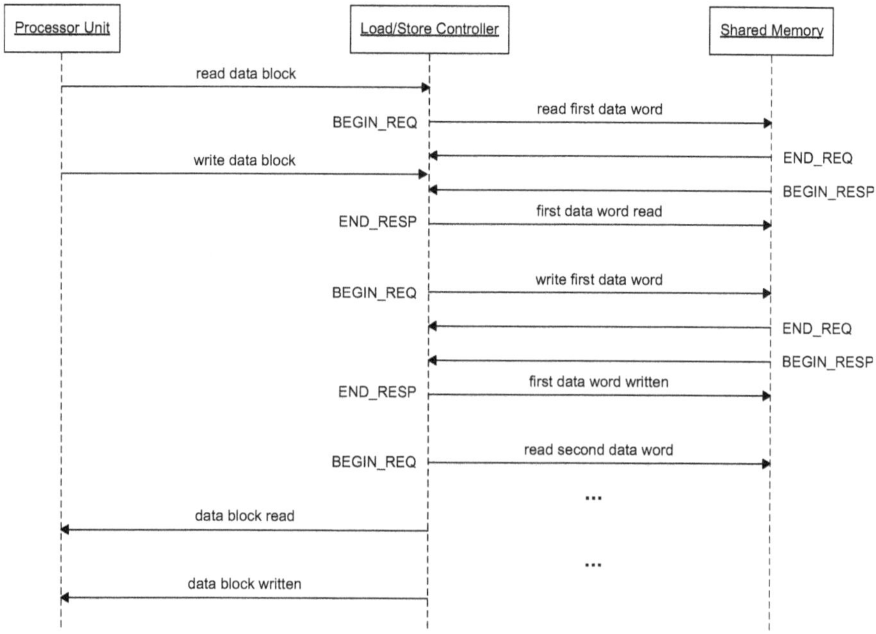

Figure 2.6: Example of a read/write data transfer sequence

transaction using the TLM2 blocking transport interface. It initiates the reading of a data block from a specified (random) address in the memory into the cache. Each block consists of a multiple of an integer word (4 bytes) for demonstration purposes. The load/store controller takes the read transaction and generates a number of burst transactions. Each burst transmits a single data word out of the data block from the memory. This behavior models the limited band width of the memory bus. The band width prevents a transfer of the data block as a whole. Every burst transaction is modeled by the TLM2 base protocol in the approximately-timed coding style using the four standard phases, i.e. BEGIN_REQ, END_REQ, BEGIN_RESP, and END_RESP (see Figure 2.6). Hence, a "real" data transfer, using handshake signaling, can be described close to reality. In the real-world design, the cached data are subsequently processed by a loaded SIMD program. As soon as processed data are available, they are written back into memory. For the sake of simplicity, the test bench initiates the write transfer after a random time and writes the cached data to a specified (random) memory address. Write transfers are equally handled like read transfers. Due to the specified single port cache, it is not possible to handle bursts in parallel. So, data transfers take place in a mutual exclusive fashion which ensures fairness. Thus the overall processing time is optimized.

Chapter 3

Early Error Detection

At an early stage errors and coding flaws in a system model can be efficiently detected by *deduction techniques* that are presented in this chapter. Figure 3.1 illustrates the necessary preconditions to apply these techniques in the proposed flow. Besides the syntactical correctness of a design description, the smallest design unit, i.e. a module, shall be available. So, a single method or function can be already analyzed. However, the expressiveness of analysis results is limited due to incompletely described functionality. Since a simulatable description is not yet available, *Static Analysis* methods are used to check particular correctness aspects of the system model. In general, hypotheses are deduced from the program code to the concrete simulation run.

Static analysis, i.e. the analysis of a program without its simultaneous execution, provides important contributions to ensure the source code quality and to detect easy-to-find errors. Code quality is guaranteed by coding standards that prevent common pitfalls and careless mistakes. Such standards usually have a language-specific structure and can be checked automatically by linting the source code. The analysis of the program code helps the designer to debug a syntactic correct program before it is executable the very first time. So, many potentially troublesome errors can be found. Static analysis is efficient, can be used with little effort, has a high degree of automation, and is applicable to arbitrary complex design descriptions. In spite of these advantages, the largest problem of static analysis is the limited precision due to the conservative approximations of reachable program states. This often results in many false warnings.

The first part of this chapter summarizes some theoretical foundations of static analysis and introduces the analysis framework REGATTA (REtarGetable FrAmework for TranslaTors and Analyzers). The framework allows the flexible and easy generation of static analyzer tools for arbitrary programming and description languages, especially for code quality assurance tasks. Generic framework components facilitate the creation of analyzer tools. A common configuration interface is based upon state machines. It allows the easy formulation of arbitrary analyses as well as the configuration of the different generic framework components.

Based upon REGATTA, the SystemC analyzer SDAS (SystemC Design Analysis System) is introduced in the second part of this chapter. Its analysis capabilities are investigated with respect to ensure the code quality of a real-

F. Rogin and R. Drechsler, *Debugging at the Electronic System Level*,
DOI 10.1007/978-90-481-9255-7_3, © Springer Science+Business Media B.V. 2010

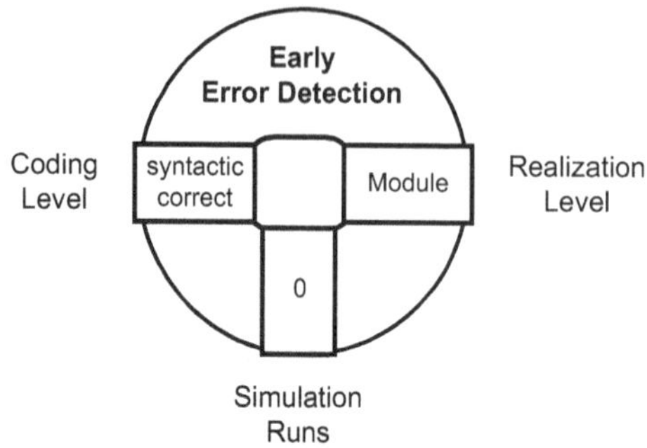

Figure 3.1: Early error detection in system models

Figure 3.2: Structure of a front-end for static analysis

world industrial SystemC verification environment. The SIMD data transfer example shows how static analysis finds a serious error prior to simulation.

1 DEDUCTION TECHNIQUES IN A NUTSHELL

Static analysis is a promising method to cope with arbitrary complex formal descriptions. It enables the designer to find many easy-to-find errors very early in the development flow. At first, this section shortly summarizes some notions as used in this work. Many data structures and algorithms have their origin in the classical compiler construction. A comprehensive discussion of static analysis and its foundations can be found in [ASU03], [WM96], [Mor98], or [Muc97]. Finally, related work, especially to check SystemC design descriptions statically, is discussed.

1.1 Preliminaries

To perform a static analysis, a compiler front-end is needed as shown in Figure 3.2. At first, the scanner performs the *Lexical Analysis*. It scans the *Input String* (the program code) and splits it into blocks of text by means of a set of regular expressions and gives them a meaning. This process is also called tokenization. In contrast to a compiler front-end, documentation and code formatting are important information while checking coding standards, e.g. the correct documentation style. Such information is not described by a context-free grammar. Thus, it is usually filtered by the lexical analysis step. However, static analysis, as used in this book, distinguishes two token classes:

- *Meaningful tokens*. Tokens of this class correspond to terminal symbols of a context-free grammar.
- *Meaningless tokens*. Those tokens summarize characters used for code formatting and documentation issues.

During *Syntactical Analysis*, the parser checks if the stream of meaningful tokens is a valid expression by means of the given context-free grammar. If a valid derivation is found, a suitable intermediate representation is created, i.e. a syntax tree.

Definition 1. *A 4-tuple $\Gamma = (V, \Sigma, R, S)$ with a finite set V of non-terminals and terminals, a finite set of terminals Σ with $\Sigma \subset V$, a finite set of grammar rules $R \subseteq (V - \Sigma) \times V^*$, and the start symbol $S \in (V - \Sigma)$ is called a context-free grammar.*

Definition 2. *Let $\Gamma = (V, \Sigma, R, S)$ be a context-free grammar. Unique parse points $\perp_i$ are added to the right-hand side of each grammar rule in R as follows: If $A \to N_1 N_2 \ldots N_k$ is the first 'A-rule' in R, so the rule with annotated parse points is given by $A \to \perp_{A11} N_1 \perp_{A12} N_2 \perp_{A13} \ldots \perp_{A1k} N_k \perp_{A1k+1}$. Π denotes the set of all parse points for Γ.*

Definition 3. *Let $\Gamma = (V, \Sigma, R, S)$ be a context-free grammar and Π the set of parse points for Γ. So, the tuple $A = (V, \Pi)$ is called the analysis alphabet of Γ.*

To process meaningless tokens during subsequent analysis steps, they are forwarded to the parser and a so called extended syntax tree is created.

Definition 4. *Let $\Gamma = (V, \Sigma, R, S)$ be a context-free grammar. A concrete syntax tree is an Extended Syntax Tree (EST),*

- *If the leaves of the tree, labeled with symbols from $\Sigma \cup \{\varepsilon\}$, are additionally labeled with meaningless token. So, the concatenation of the leaves from left to right completely gives the parsed code including all printable characters.*
- *If location information (row, column, filename) are attached at each token node.*
- *If an arbitrary inner tree node is marked with A and all its child nodes are marked from left to right with $N_1, N_2 \ldots, N_k$, so $A \to N_1 N_2 \ldots N_k$*

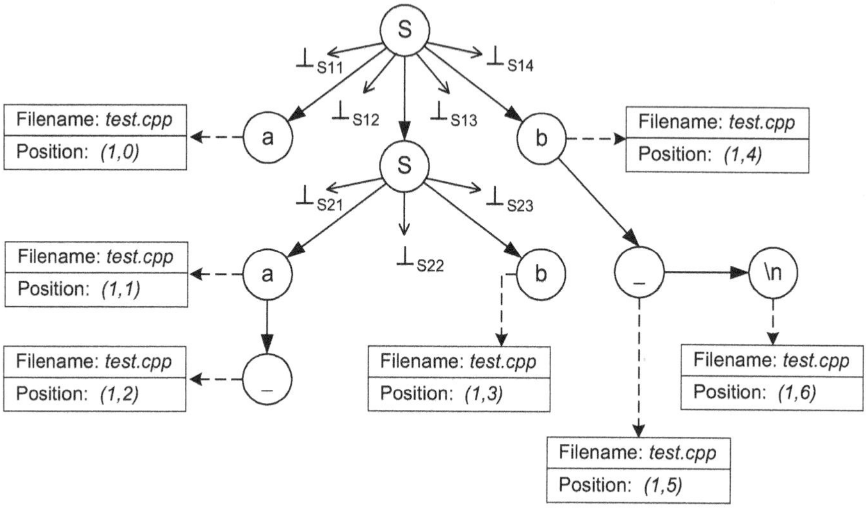

Figure 3.3: EST for the word "aa_bb_\n"

is a grammar rule in R. A is annotated according to Definition 2. For each parse point a tree node is created and labeled accordingly.

An EST stores all parsed information for a later analysis. Here, the use of parse points eases the formulation of analyses.

Example 2. *A grammar* Γ *for the context-free language* $L = \{a^n b^n, n \geq 1\}$ *is defined as follows:* $\Gamma = \{\{S, a, b\}, \{a, b\}, \{S \rightarrow aSb, S \rightarrow ab\}, S\}$. *A word of L is for instance "aa_bb_\n". The characters '_' and '\n' represent code formatting information (spaces, linefeed) that are not processed normally. According to Definition 4, Figure 3.3 depicts the corresponding EST.*

Most methods for syntactical analysis distinguish between two parsing techniques: *top-down parsing* and *bottom-up parsing*. Both parsing techniques read the input (tokens) from left to right. Since the manual creation of a parser could be expensive, specific language classes allow to create a parser by using a parser generator automatically. Therefore, the user has to provide only an appropriate context-free grammar. Several tools generate a parser for LL(k) and LALR(1) grammars [ASU03]. These classes of grammars recognize most programming and description languages.

Given an EST, a large number of lexic- and syntax-driven analyses can be already implemented. However, semantic analyses are based upon processed data structures. A common data structure is the symbol table. It holds a set of attributes for each recognized identifier, e.g. the point of declaration, the type, or scope level. A further set of analyses is based on control and data flow information of the analyzed program.

Definition 5. *A basic block B in a program is a sequence of consecutive statements with unique entry and exit points. That means, there is exactly one way to enter and leave the block.*

Definition 6. *A 4-tuple* $G = (N, E, s, e)$ *with* N *a finite set of nodes,* $E \subseteq N \times N$ *a finite set of directed edges, a start node* $s \in N$*, and an end node* $e \in N$ *is called Control Flow Graph (CFG). An additional labeling function* $f_L : N \rightarrow \Phi$ *assigns each node a part of the program.* Φ *denotes the set of fragments of the created syntax tree.*

Each basic block becomes a node in *G*. The edges define the flow of control between blocks. A CFG defines all possible execution paths in the analyzed program. An annotation function $f_A : N \rightarrow F$ assigns a particular data flow information *F* to each node in *G*, e.g. variable accesses. Based on this information, a *Data Flow Analysis* (DFA) can determine which data reach a particular program point.

To facilitate the generation of tools for static analysis, an *object-oriented software framework* could define core architecture and basic functionality of such tools.

Definition 7. *Johnson defines a software framework as "a skeleton of an application that can be customized by an application developer" [Joh97].*

According to Definition 7, a software framework is a semi-finished software component. It facilitates the development of a software product by providing an API and pre-fabricated problem solutions for a particular application domain and certain use cases. A framework summarizes the commons of such a domain. It defines important design decisions for the software architecture of applications based on it. The development of a framework requires a solid and in-depth analysis of the application domain to anticipate many future use cases. In contrast to a library, especially the design decisions of the framework architecture are reused. The principle of the *inversion of control* is a distinctive important characteristic of a software framework. This design principle is also known as the *Hollywood-Principle* "don't call us, we will call you". Here, the control remains at the framework side where generic code controls the execution of problem-specific code extensions.

1.2 Related Work

In [HR06] static code analysis is used to calculate classical SW-oriented structural metrics of SystemC designs, e.g. the depth of the class hierarchy. A set of HW-based functional metrics is also calculated, e.g. to estimate hardware implementation decisions by considering the average execution time. Agosta et al. [ABS03] present a similar approach. Here, metrics are computed on basis of a formalized, abstract TLM description. Both work extract metrics especially to support design exploration. The generic functionality and configurability of our analysis framework REGATTA would enable the easy implementation of similar analyses.

Several work statically analyze SystemC descriptions. Siebenborn et al. [SBR02] determine temporal system properties by a timing analysis of parallel communicating processes, e.g. the worst-case response time. Blanc et al. [BKS08] generate a static scheduler from an extracted formal model. The characteristic of most approaches is the usage of static analysis to create a formal model of a SystemC description to support formal verification. In fact, the generation of meaningful and detailed formal models is restricted to a subset of SystemC. So, the tools have limits to handle arbitrary complex SystemC designs (see Section 2.4 on page 21). Rather, our flexible framework approach supports the easy generation of tools for an analysis of corporate coding standards.

Due to the nature of SystemC being a C++ class library, any verification tool developed for C/C++ static program analysis is applicable. In this domain many powerful tools and approaches have been published over the years. In the following some representative work are detailed.

The Saturn program analysis system [AB+07] translates a C program into a set of relations. Saturn uses constraints and a logic programming language to express analysis algorithms. A similar approach is proposed by Lam et al. [LW+05]. The presented analysis framework stores all program information as relations in a deductive database. This database uses BDDs for storing while the database query language Datalog [Ull89] is used to formulate analyses. BDDs allow to handle the large set of program contexts very efficiently. Further tools use BDDs as analysis back-ends such as [BNL05]. SLAM is a toolkit [BR02] that statically analyzes a C program determining whether or not it follows defined API usage rules. SLAM validates program behavior and detects errors performing a reachability analysis of a generated Boolean program abstraction. Beyer et al. [BHJM07] introduce the model checker BLAST for C programs. It checks temporal safety properties of C program based on a lazy predicate abstraction and interpolation-based predicate discovery of the program state space. Godefroid [God97] presents the VeriSoft tool. This tool systematically searches the state space of a concurrent C/C++

program to verify defined properties. In contrast to the aforementioned model checkers, VeriSoft does not rely on any additional abstraction. Its focus lies on finding bugs rather than to prove program correctness. A pragmatic approach to find many serious bugs in C/C++ programs is proposed by Hallem et al. [HCXE02]. The user writes compiler extensions using the provided language *metal*. These extensions are executed by an analysis engine using a context-sensitive, interprocedural analysis. There, the analysis applies each extension to a constructed CFG. A similar project is ESP [DLS02] which also provides a state machine language to formulate analyses. Other approaches base upon user annotations, e.g. [Det96], [FTA02]. Generally, annotations pose a significant drawback due to the additional effort for invasive code modifications.

Apart from a few approaches, e.g. [HCXE02], [DLS02], [AB+07], many tools, e.g. [BHJM07], [God97], [LW+05], do not scale very well for large systems due to the used underlying data structures or approaches. Program sizes are usually limited to several 10 K lines of code. However, actual tools produce sound and reasonably precise analysis results [AB+07], [LW+05], [BHJM07]. In contrast, practical experiments have proven the feasibility of our SystemC analyzer SDAS to analyze large code bases very fast. The disadvantage of SDAS is the missing precision due to the lack of a pointer alias analysis and an interprocedural DFA. On the other hand, the underlying REGATTA framework was mainly developed to ensure code quality which includes simple lexical as well as advanced semantic checks. Especially, the processing of meaningless tokens is a unique feature of all REGATTA-based tools. The analysis configuration and specification approach in REGATTA is similar to *metal* [HCXE02] and ESP [DLS02] where state machines are used to describe analyses. Abstraction-based model checkers, such as BLAST, SLAM, or Saturn, are working on a previously computed abstraction of the state space of a program. This abstraction process is usually connected with a loss of precision. SDAS performs many analyses on-the-fly, i.e. during parsing the input, and relies on an exhaustive search of the concrete state space only for DFA-based checks. The drawback of C/C++ tools is the missing support to process SystemC statements directly. Hence, SystemC-specific analyses can be realized only with difficulty.

The concept of a framework providing generic analysis functionality is proposed by various work. Fechete et al. [FKB08] introduce a framework that creates CFGs for arbitrary Ada programs. Indus is a framework for the analysis and slicing of concurrent Java programs which provides a scripting interface to access collected analysis information [RH06]. The AJANA framework analyzes AspectJ programs to provide a dataflow representation that is the basis for an interprocedural dataflow analyses [XR08]. This work focuses on specific analysis problems introduced by aspect-oriented language features. The difference between these approaches and REGATTA is the limitation to a single programming language although the tools sometimes enable more sophisticated analyses. Moreover, REGATTA was the basis for various framework-based tools applied on different languages such as Verilog [HR08], VHDL-AMS [RFSH05], or even SystemC. The analyzer generator PAG [Mar98] shares with REGATTA the general objective of providing a generic framework for analyses of various programming languages. PAG bases on abstract interpretation and the specification of data flow analyses using a high-level functional input language. There, it relies on CFG construction by a third-party compiler front-end. In contrast, REGATTA integrates its own front-end and performs analyses using a concrete syntax tree. So, it supports the analysis of coding standards very well whereas PAG is restricted to dataflow-based checks.

There exist several (commercial) SystemC design and verification environments. They provide various analysis, debugging, and visualization capabilities such as CoWare's Platform Architect [CoWare], or Summit's Design Environment for SystemC Vista [Summit]. All these environments are adapted to a specific application domain [CoWare], or base upon an extended and partially proprietary simulation kernel [CoWare], [Summit]. In spite of the comprehensive tool features, quality assurance by static code is only partially or not supported. There, only AccurateC [Actis] directly checks the code quality of SystemC descriptions.

Summarized, static analysis techniques are state-of-the-art in software engineering and hardware design but in system design their application is still limited yet, especially in the SystemC domain.

2 STATIC ANALYSIS FRAMEWORK

First, this section summarizes important requirements for the development of a framework for static analysis. Subsequently, the general architecture of the REGATTA framework and the pre-defined generic components are detailed. Next, a flexible configuration approach is introduced. This approach eases the adaptation of generic framework components to a concrete analyzed language. Moreover, this approach is flexible enough to formulate arbitrary analyses. Finally, a rating of the developed framework is given.

2.1 Requirements

To facilitate the generation of arbitrary static analyzer tools, the development of the REGATTA framework is based on certain requirements.

- *Flexibility.* A suitable front-end for lexical and syntactical analysis should be created easily and automatically. Therefore, a LR(k) or a LL(k) grammar for the desired language is needed. Hence, the approach shall not depend on existing compiler front-ends. To obtain the required flexibility, language processing (front-end) and analysis functionality (back-end) should be decoupled from each other.
- *Adaptability.* The framework architecture should permit an easy adjustment of REGATTA to various languages and analyses. An user-friendly and powerful standardized configuration interface shall allow the user to formulate new analyses and to adapt predefined generic framework functionality.
- *Integration.* The framework shall be able to supplement existing (commercial) analysis tools. Proper framework components should support the particular integration, e.g. by generating analysis results in a specific output format.

The REGATTA framework should support various use cases:

- *Ensuring coding standards.* The most important use case is the analysis of a particular coding standard. Possible analyses concern for instance the compliance check of naming conventions, the check of a correct code documentation, or dataflow analyses.
- *Structural analysis.* REGATTA shall permit a structural analysis of the source code. Structural analysis can be used for instance to measure the design or code quality of the analyzed description.
- *Processing of analysis data.* Apart from code analyses, the framework should allow the further processing of collected and analyzed data, e.g. to generate an HTML documentation from a given design description.

2.2 General Architecture

Figure 3.4 illustrates the general architecture of REGATTA. The *Analysis Manager* is the central controlling unit. It evaluates the user input, creates and initializes all framework components, and contains the main control loop for the analysis flow. The *Front-end* divides into a *Scanner* and a *Parser* per-

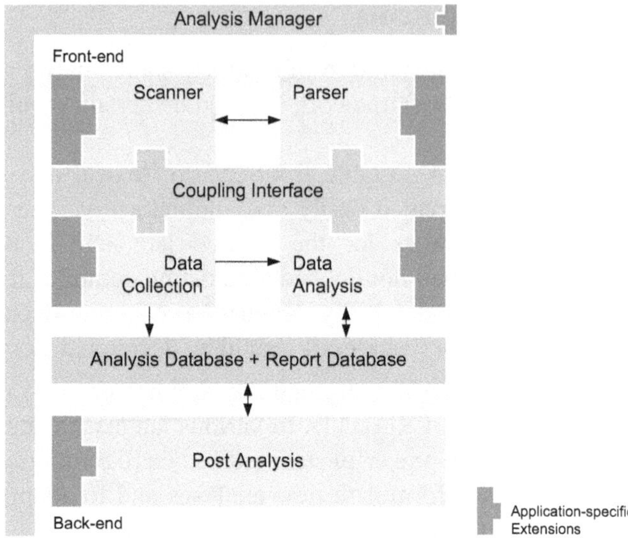

Figure 3.4: General architecture of REGATTA

forming the lexical and the syntactical analysis. A concrete front-end is generated by a parser generator automatically. Therefore, the user has to supply a proper context-free grammar of the target language. REGATTA supports two generator toolsets which maximize the framework flexibility in supporting a wide range of programming and description languages:

- Lex/Yacc [LMB92] or flex/bison [Don92] creates for a given context-free LALR(1) grammar an LALR(1) parser and a corresponding table-driven scanner.
- ANTLR [Parr07] generates parser and scanner using a provided LL(k) grammar.

The *Back-end* comprises all components to analyze the given input description by means of a created EST (Definition 4). Each analysis task can be split into a *Data Collection* and a subsequent *Data Analysis* phase. Either data are collected globally for various analyses or they are gathered task-specific locally. The subsequent data analysis phase evaluates the collected data according to the desired analysis and writes possible results into the database. Two analysis types are distinguished: The first one performs an on-line analysis that means executes the whole check during tree walking the EST. The second analysis type, also *Post Analysis*, is performed after tree walking is done. It works directly on data that have been collected into the database before. The framework distinguishes two types of databases, i.e. the *Analysis Database* (DB_A) and the *Report Database* (DB_R):

- The DB_A stores all analysis-specific data either globally for many different analyses, or locally for specific checks.
- The DB_R holds all results of performed analyses such as error reports in case of violated checks or common logging messages from framework components.

To reach the demanded framework flexibility, front-end and back-end are separated by a *Coupling Interface*. This interface facilitates the transparent adaptation of the framework core according to the concrete formal language (front-end) and the desired analyses (back-end). Framework configurability is enabled by application-specific extension points. Such a point is provided by using object-oriented dcsign-patterns [GHJV95] or the common configuration approach by using *FSM-Described Configuration* (FDC) specifications (see Section 2.4 on page 51).

2.3 Generic Framework Components

Generic framework components ease the adaptation of REGATTA to various analyses. The generic symbol table accelerates the implementation of a concrete symbol table. The DFA component supports the development of dataflow-based analyses, e.g. to detect dataflow anomalies. Thus, potentially troublesome errors can be found very easily. Finally, the structural analysis component facilitates the detection of structural patterns in an analyzed description. Structural analysis allows to estimate the code quality in terms of maintainability and extensibility for instance.

2.3.1 Generic Symbol Table

Many advanced analyses base upon the processing of symbols such as variables, functions, or classes. Possible questions in case of an analysis of coding standards could be for instance:

- Does a module name contain the name of its top module?
- Does a specific method lay inside each class?
- Does the depth of the class inheritance tree exceed a maximum value?

To answer such questions, a symbol table is used. The symbol table saves information about the symbols defined in a program, e.g. the validity of a variable in the current scope, the required memory of a symbol type, or the number and types of arguments of a method. A compiler creates a symbol table for semantical analyses and to generate target code.

REGATTA supports symbol processing by providing a generic symbol table component that has to be configured according to the analyzed language. Important requirements of REGATTA are the adaptability and the inherent language independent character (see Section 2.1 on page 41). This is also reflected by the design requirements used to implement the generic symbol table:

- *Hierarchical design.* If the analyzed language supports the creation of multiple scopes that can be interleaved, the symbol table shall be able to model this fact.
- *Distinction between symbols and types.* In a typed language, each symbol is of a defined type. According to the language, there are predefined and eventually user-defined types available.
- *Language independence.* The symbol table should provide generic data types that hold common information about symbols and types of typical programming and description languages.
- *Configurability.* An interface shall allow the configuration of the generic symbol table according to the analyzed language.
- *Extensibility.* Unsupported symbols or types shall be added easily into the existing design structure. Moreover, language-specific scope rules should be efficiently integrated.
- *Efficiency.* Retrieving a symbol and its attributes shall be possible very efficiently to allow fast symbol access, and thus a fast analysis.

REGATTA implements a generic symbol table by following the aforementioned requirements. Figure 3.5 depicts the class hierarchy:

- *Hierarchical design.* Each symbol table holds a pointer to its super table which is the table of the enclosing father scope. Hence, a scope hierarchy can be created and traversed easily. The symbol table provides three predefined data types that open a new scope, i.e. DBI_Subtype, DBI_GroupType, and DBI_MethodType.
- *Distinction between symbols and types.* The symbol table stores symbol table objects (class DBI_SymTabObj) which is a common base class for all symbols and types. A symbol (class DBI_Symbol) denotes a parsed identifier that is of a specific type. A type (class DBI_Type) encapsulates all type-relevant attributes of a symbol. Separating symbols and types enables the reusability of the same type object for different symbols.

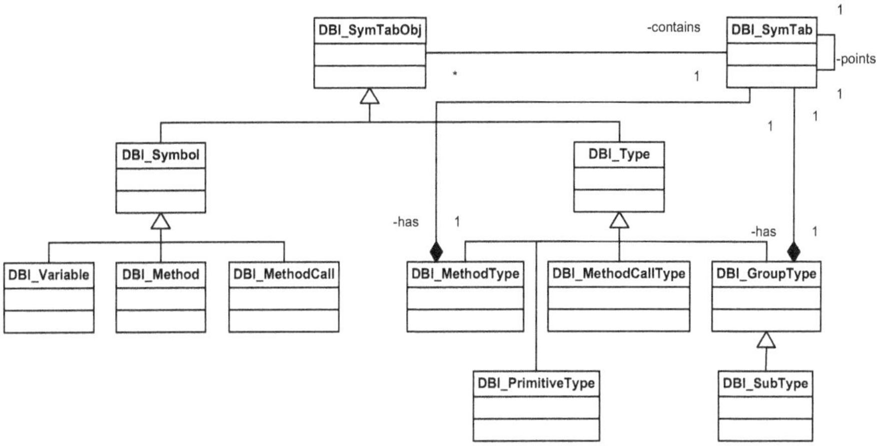

Figure 3.5: Extract of the generic symbol table class hierarchy

- *Language independence.* As shown in Figure 3.5, REGATTA predefines several symbol and type classes. These classes provide symbol handling for a wide range of languages. The focus lies especially in the support of programming and specification languages such as SystemC, C/C++, or Java. However, the symbol table architecture can be also mapped to HDLs such as Verilog or VHDL.
- *Configurability.* FDC specifications (see Section 2.4 on page 51) control the various construction steps for a specific symbol or type object. Hence, a single and unique configuration interface for symbol table construction is provided.
- *Extensibility.* The common base for symbols and types (class DBI_SymTabObj) makes it easy to extend the symbol hierarchy. The particular scope rules of the analyzed language define the visibility of symbols. To adapt the corresponding symbol lookup algorithm, the *Strategy* design pattern is used [GHJV95]. This pattern eases the implementation of new lookup strategies.
- *Efficiency.* Symbol table objects are stored in the related symbol table object using a hash map. Hence, insertion and retrieval of object is possible in constant time, i.e. O(1).

2.3.2 Generic Dataflow Analysis

Dataflow-based analyses have their origin in the classical compiler construction. They are used for code optimization and error checking, e.g. reporting unused variables, or the correct call of a method. REGATTA applies such analyses to detect data flow anomalies. There, an anomaly is a deviation of a property from its specified behavior. The generic DFA component divides into two parts:

- *CFG construction.* This part includes the construction of a CFG G (Definition 6), and its subsequent annotation with data flow information using the annotation function f_A.
- *DFA algorithms.* By means of an annotated CFG G, DFA algorithms solve data flow equations in order to locate errors in the code.

During the first step a CFG is constructed for each recognized syntactical block, e.g. a function which allows to perform an *intraprocedural analysis*. Since REGATTA is aimed to be language independent, data structures have to fit for many different but typical language constructs. For that case, so called *control flow patterns* are proposed. These patterns support CFG construction (see Figure 3.6):

- *Sequence.* A sequence is just a sequence of two blocks. This pattern is an intermediate step to model all other structures. After construction of the CFG, there should be no sequence pattern anymore.
- *Fork.* A fork marks a point in the control flow where several (at least two) ways (branches) are possible in the control flow. In contrast to a breakable fork, no more than one branch of the fork can be used during program execution. An example for such a fork is the `if-then-else` statement in SystemC.
- *Breakable fork.* A breakable fork allows to break the execution of a branch and to switch to another branch of the same fork. The branches are executed in sequence as long as no command for breaking up the actual fork is found. An example for this kind of fork is the `switch` statement in SystemC where the `break` statement controls branch execution.
- *Loop.* A loop is a point in the control flow where a set of basic blocks is possibly executed several times. The loop is introduced by a start block, including the loop-condition, a set of looping blocks, representing the control flow within the loop, and an end block. There, the end block points to the first position after the loop. A further loop variant executes the loop body at least one time and checks the loop-

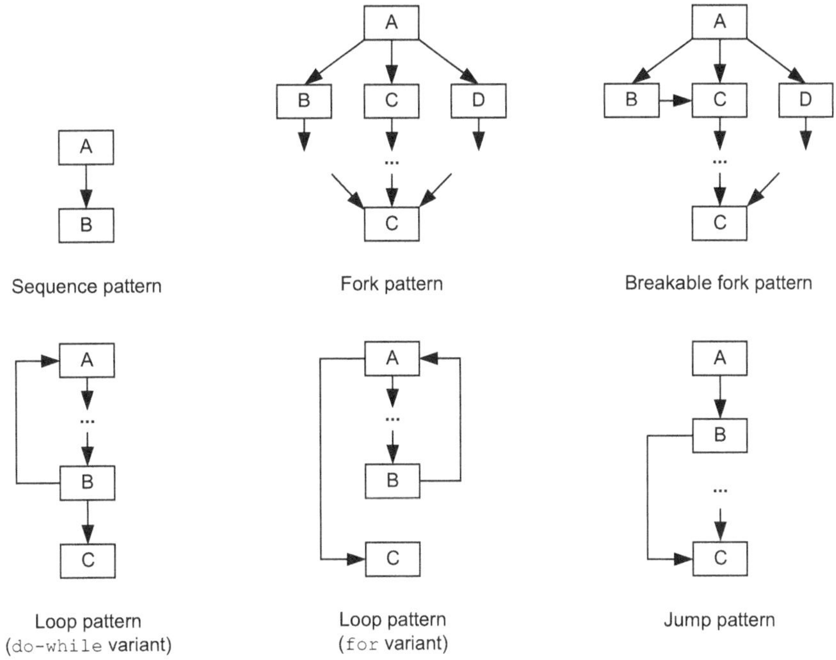

Figure 3.6: Control flow patterns recognized in REGATTA

condition at the end. Examples for both loop variants are the `for` and the `do-while` statements in SystemC.

- *Jump.* A jump connects a block to another block somewhere else. It can be a backward or a forward jump. Throughout the local creation of a CFG, a jump target could not yet be known during parsing. So, after parsing, all jumps are "connected" to the appropriate target blocks. There, it is assumed that the jump target can be deterministically located. An example for a jump is a method call in SystemC.

The control flow patterns are implemented using abstract FDC specifications that are tailored to the concrete analyzed language (see Section 2.4 on page 51). In parallel to the creation of a CFG G, the annotation function f_A assigns each node in G, i.e. each basic block, a particular data flow information. Here, variable accesses are annotated where four access classes are distinguished, i.e. *read* (also *referenced*), *written* (also *defined*), *used* (also *defined*), and *declared.* A particularity is the *used* access. Sometimes, it is undecidable whether a variable is read or written, e.g. the (indirect) modification of a variable content using pointer arithmetic. Here, a sound analysis has to assume a read as well as a write access.

```
for each basic block B in G do begin
    for all definitions D of variable v in B do begin
        put the last definition D of v into gen[B]
    end
end
```

Figure 3.7: Algorithm to determine the *gen* sets for each basic block

```
for each basic block B in G do begin
    for all definitions D of variable v in gen[B] do begin
        get all (global) definitions of v
        put all (global) definitions except D into kill[B]
    end
end
```

Figure 3.8: Algorithm to determine the *kill* sets for each basic block

The solution of DFA problems is facilitated by an implementation frame. This frame helps the programmer to solve data flow equations as described by Aho et al. [ASU03]. The basis of these equations are variable definitions that reach a program point. There, it is assumed that each path in *G* can be a real execution path (*conservative approximation*). A program point can generate and kill variable definitions. The definition of a variable *v*, that reaches the end of a basic block *B*, is generated by this block (see Figure 3.7). Simultaneously, this definition kills all other (previous) definitions of *v* (see Figure 3.8). Using the algorithms shown in Figures 3.7 and 3.8, *gen* and *kill* sets of every basic block in *G* are calculated.

Example 3. *Figure 3.9 shows for a CFG on the left side the corresponding gen and kill sets. In* B_2 *variable c gets a new value for instance which kills the definition of c in* B_1 *simultaneously.*

After calculating *gen* and *kill* sets for each basic block, the reaching definitions in terms of *in* and *out* sets for a blocks *B* can be calculated. The *in*[*B*] set contains all definitions that reach the beginning of *B*. The set *out*[*B*] contains the definitions which leave *B*. Figure 3.10 illustrates an algorithm that calculates all definitions that reach the basic blocks of a created CFG, similar to the procedure presented by Aho et al. [ASU03]. The implementation of this algorithm can be used as a template to add further dataflow anomaly analyses. The main difference to the algorithm, as given by Aho et al., is the calculation of

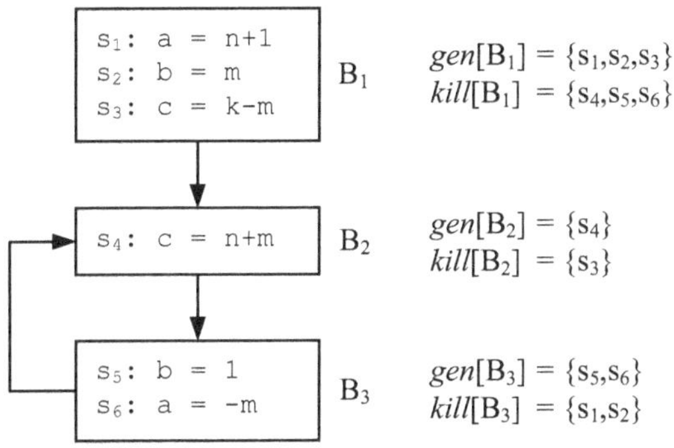

Figure 3.9: *gen* and *kill* sets in an example graph

```
1:  for each basic block B in G do out[B] = gen[B]
2:  change = true
3:  while change do begin
4:      change = false
5:      for each block B do begin
6:          in[B] =    ∩      out[P]   // pre(B): set of predecessors of B
                    P ∈ pre(B)
7:          oldout[B] = out[B]
8:          out[B] = gen[B] ∪ (in[B] – kill[B])
9:          if out[B] != oldout[B] then change = true
10:     end
11: end
```

Figure 3.10: Algorithm to detect undefined variables (according to Aho et al. [ASU03])

definitions that reach *B* (line 6). A compiler optimization must be safe. So, the set *in*[*B*] is calculated by using the union over all *out* sets of all direct predecessor blocks. Hence, no variable definition is missed (conservative approximation). However, static code analysis checks whether there is at least one path a variable is not defined, so that the intersection is calculated instead.

2.3.3 Generic Structural Analysis

REGATTA supports a structural analysis of design descriptions using the open source tool *CrocoPat* from Beyer et al. [BNL05]. *CrocoPat* is a tool for the efficient manipulation of relations of arbitrary arity. It was developed to find structural patterns in graph models of software systems which includes different use cases:

- detection of particular structural design patterns, e.g. as presented in [GHJV95],
- extraction of scenarios from the source code, e.g. specific function call sequences,
- detection of potential design problems, or
- study of the impact of code changes in software systems.

To use *CrocoPat*, a REGATTA component is able to generate user-defined, arbitrary relations. These relations describe dependencies between different elements in the analyzed description, e.g. classes, variables, methods. Based on a set of generated relations, a *CrocoPat* query implements an analysis. This procedure includes several benefits:

- *Ease of adaptation. CrocoPat*-based checks are evaluated before the analysis process starts. Hence, a change of the check does not require a recompilation of any source code.
- *Ease of use.* The *CrocoPat* programming language bases upon first-order predicate calculus which is well known, reasonably simple, precise, and powerful. So, check implementation is a fairly simple task.
- *Efficiency. CrocoPat* handles relations of arbitrary arity which allows to perform complex analyses.

Structural analysis in REGATTA is composed of two phases:

- *Initialization.* Before the input sources are parsed, all selected *CrocoPat* queries are read into the DB_A. There, the XML configuration interface of REGATTA is used. A query is introduced by the XML tag *Query*.
- *Analysis.* After the input program is analyzed, processed and the demanded relations are generated, *CrocoPat* evaluates the queries. An input relation consists of the relation name and an arbitrary number of elements describing the desired information. Finally, analysis results are written into the DB_R.

Figure 3.11(b) presents a *CrocoPat* query that counts the number of *Composite* design pattern instances [GHJV95] found in the analyzed description.

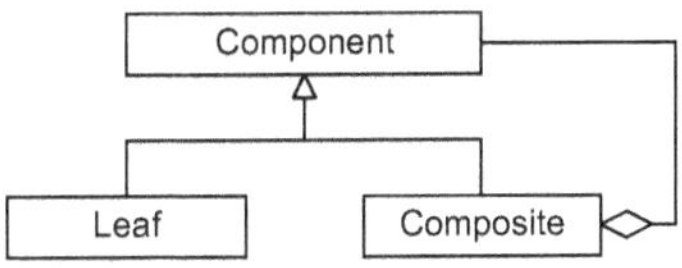

(a) Structure of Composite pattern

```
Composite(Component, Composite, Leaf) :=
    INHERIT(Composite, Component) &
    CONTAIN(Composite, Component) &
    INHERIT(Leaf, Component) &
    !CONTAIN(Leaf, Component);

IF (#(Composite(Component,_,_)) > 0) {
    PRINT "#", #(Composite(Component,_,_));
}
```

(b) *CrocoPat* query

Figure 3.11: Detection of the Composite design pattern using *CrocoPat*

The basis for that query are three relation types that describe typical relations in object-oriented programming languages:

- INHERIT A B means type A inherits from type B
- CONTAIN A B means type A contains type B, and
- CALL A B means type A calls type B.

2.4 Configuration and Adaptation

Practical experiences with different REGATTA-based analyzers have identified a large number of always recurring analysis problems, e.g. the check of the same coding guideline, or the construction of the symbol table. Hence, a simple, reusable, adaptable, and language-independent approach for the specification of analyses in REGATTA should be available. The approach shall met the following requirements:

- *Implementation of analyses.* A single point in the framework allows for an abstract, concise, and easy specification of analyses.
- *Tailoring to the analyzed language.* An abstract analysis specification should be tailored to the analyzed language without any need for implementation changes.

2.4.1 Implementation of Analyses

Finite-State Machines (FSM) are a suitable and flexible abstraction to provide a concise, easy to use, and familiar specification facility. Hence, code analyses as well as generic framework components can be described (see Section 2.3 on page 43). The notation of an FSM-based specification approach in terms of FDCs is geared to the syntax of the State Machine Compiler [Rapp08].

Definition 8. *Let* $\Gamma = (V, \Sigma, R, S)$ *be a context-free grammar and A the analysis alphabet of* Γ. *An FDC specification is a 12-tuple* $Z = (Q, \Sigma, F_{act}, E, \delta, \lambda, F_{entry}, F_{exit}, \gamma_{entry}, \gamma_{exit}, q, Q_{err})$ *with*

- Q: *the finite number of states representing the analysis steps,*
- Σ : *the input alphabet defined over A,*
- F_{act} : *the finite set of functions describing actions,*
- *E: the finite set of C++ Boolean expressions to guard transitions,*
- δ : *the transition function* $\delta : Q \times \Sigma \times E \rightarrow Q$
- λ : *the output function* $\lambda : Q \times \Sigma \times E \rightarrow F_{action}$,
- F_{entry} : *the finite set of entry functions* $F_{entry} = F_{entry} \cup \{\varepsilon\}$,
- F_{exit} : *the finite set of exit functions* $F_{exit} = F_{exit} \cup \{\varepsilon\}$,
- γ_{entry} : *the function* $\gamma_{entry} : Q \rightarrow F_{entry}$ *assigning each state in Q an entry function,*
- γ_{exit} : *the function* $\gamma_{exit} : Q \rightarrow F_{exit}$ *assigning each state in Q an exit function,*
- q : *the initial start state* $q \in Q$, *and*
- Q_{err} : *the finite set of error states* $Q_{err} \subset Q$.

A coding standard is ensured by a set of coding checks implementing particular coding guidelines. The outcome of an FDC-based check is defined by two different path types.

Definition 9. *Let* $Z = (Q, \Sigma, F_{act}, E, \delta, \lambda, F_{entry}, F_{exit}, \gamma_{entry}, \gamma_{exit}, q, Q_{err})$ *be an FDC specification. A sequence* $\pi_{valid} = q, \ldots, q_{end}$ *of states is a valid path from q to* q_{end} *in Z, if for a sequence* ω *with* $\omega \in \Sigma^*$, $\hat{\delta}(q, \omega, E_1) = q_{end}$ *with* $E_1 \subseteq E$, $q_{end} \notin Q_{err}$ *applies.* Π_{valid} *is the set of all valid paths.*

Definition 10. *Let* $Z = (Q, \Sigma, F_{act}, E, \delta, \lambda, F_{entry}, F_{exit}, \gamma_{entry}, \gamma_{exit}, q, Q_{err})$ *be an FDC specification. A sequence* $\pi_{invalid} = q, \ldots, q_{end}$ *of states is an invalid path from q to* q_{end} *in Z, if for a sequence* ω *with* $\omega \in \Sigma^*$, $\hat{\delta}(q, \omega, E_1) = q_{end}$ *with* $E_1 \subseteq E$, $q_{end} \in Q_{err}$ *applies.* $\Pi_{invalid}$ *is the set of all invalid paths.*

A coding guideline is violated if there is a sequence ω of program statements, so that Definition 10 applies. The analyzed program satisfies the coding guideline if no path $(q, \ldots, q_{end})$ for the program exists, so that $(q, \ldots, q_{end}) \in \Pi_{invalid}$. FDC specifications are also used to describe generic framework components. In that case, there is no explicit error state in *Z*, i.e.

```
fdc_spec       ::= {fdc_desc}+
fdc_desc       ::= t_FDC ident '{' {require_decl}* {import_decl}* {var_decl}*
                               [prior_decl] [init_decl] {func_decl}*
                               {state}+ '}'
require_decl   ::= t_REQUIRE <include_path>
import_decl    ::= t_IMPORT type ident ';'
var_decl       ::= t_VAR var_type ident ';'
prior_decl     ::= t_PRIORITY number ';'
init_decl      ::= t_INIT '{' action '}'
func_decl      ::= t_FUNCTION ident '{' action '}'
state          ::= state_name [entry] [exit] '{' {trans}+ '}'
trans          ::= proxy [ '[' guard ']' ] state_name [ '{' action '}' ]
proxy          ::= ident
state_name     ::= ident | t_NIL
entry          ::= t_ENTRY '{' action '}'
exit           ::= t_EXIT '{' action '}'
type           ::= t_INT | t_STRING | t_DOUBLE | t_BOOL
ident          ::= [a-zA-Z_][a-zA-Z0-9_]+
var_type       ::= // any correct C++ type
action         ::= // correct C++ code + Regatta API
guard          ::= // correct C++ Boolean expression
```

Figure 3.12: EBNF syntax of the FDC language

$Q_{err} = \varnothing$. Rather, the FDC automaton is reset to q in case of an erroneous situation.

2.4.2 Tailoring to the Analyzed Language

To create an abstract FDC specification, all language-specific information have to be removed. Mainly, this concerns the symbols in Σ which are replaced by so called *proxy symbols*. In case of a concrete analyzer, these proxy symbols are assigned elements from the analysis alphabet A of the particular given grammar. The check functionality is abstracted by using special *imported variables*. Variable values are set by the designer accordingly to the concrete analysis, e.g. to define the maximum limit for the number of allowed function arguments. Practical experiences have shown that the primitive types *bool*, *string*, *integer*, and *double* are sufficient to support a large set of analyses.

To tailor an abstract FDC specification to a concrete language such as SystemC, an XML configuration interface is provided. This interface is generated automatically when an FDC specification is compiled. It provides XML tags for each proxy symbol and the defined imported variables. The configuration interface facilitates an easy and quick adaptation of abstractly formulated analyses.

Based upon Definition 8, Figure 3.12 specifies the syntax of FDC specifications in *Extended Backus Naur Form* (EBNF) [ASU03]. The action code of F_{act}, F_{entry}, and F_{exit} have to be correct C++ code where the framework

```
fdc ComponentCnt {

  import int max_cnt;    // maximum allowed count
  import string pat;     // filename regular expression
  var int cur_cnt;       // current component counter
  var int active;        // true if check is active

  function check_file
  {
    // match filename with given regular expression
  }

  function create_msg
  {
    // create and store an error message
  }

  // FSM specification
  Start {
    PK_INIT Counting { cur_cnt = 0; check_file(); }
  }

  Counting
  {
    PK_CNT [ active && ++cur_cnt > max_cnt ]
                          Error { create_msg(); }
    PK_RESET Start
  }

  Error
  {
    PK_CNT nil { create_msg(); }
    PK_RESET Start
  }
}
```

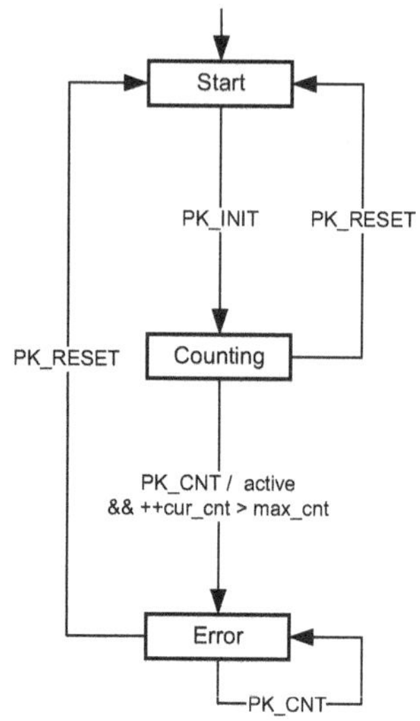

Figure 3.13: Abstract FDC specification counting arbitrary language elements

enables the full access to the REGATTA API. Appendix A gives an informal semantics of the FDC language.

Example 4. *Figure 3.13 shows an abstract FDC specification named "ComponentCnt" that counts the number of a user-defined language element. If a maximum count is reached the check creates an error message (function "create_msg"). The user has to configure the count limit "max_cnt" and a regular expression "pat" that constrains the analysis to particular files (function "check_file"). Moreover, the user has to assign the proxy symbols elements from the analysis alphabet A of the given grammar. The transition "Start-Counting" initiates the analysis. It will be released by symbols from A which are defined for the proxy symbol PK_INIT. Each occurrence of a language element is counted provided that a matching file is examined. The counted language element is notified by further symbols from A assigned to the symbol PK_CNT. If the current count exceeds the maximum count (guard condition), an error message is created and the state machine enters the "Error" state. In this state subsequent violations are reported, as well. Two reset transitions restart the check.*

2.5 Approach Rating

The following section summarizes the benefits and limitations of static analysis techniques in general and especially regarding the REGATTA framework.

2.5.1 General Benefits

A static analysis of program code has a number of advantages:

- *Scalability.* Static analysis usually scales very well with complex design descriptions, where it detects many (easy-to-find) errors.
- *Development costs.* The overall development costs are reduced due to an early detection of errors just before runtime.
- *Objectivity.* The automated analysis allows an objective estimation of code quality. Static analysis supports code reviews in highlighting confusing or problematic constructs.
- *Ease of use.* Appropriate analysis tools can be used with little effort, require only few user interaction, and are usually fully automated. Moreover, these tools directly point to the fault location which ease debugging.
- *Code quality.* The validation of a corporate coding standard raises designer's awareness for typical mistakes and pitfalls. Moreover, code readability, maintainability, reusability, and portability are enhanced.

Several work evaluate the effectiveness and accuracy of static analysis mainly in the software domain, e.g. [NW+04], [NB05], [ZW+06], [AP+07]. The efficiency of a tool-supported approach that helps the developer in preparing a manual code review is explored by Nagappan et al. in [NW+04]. Nagappan and Ball [NB05] show that defect density of static analysis predicts the pre-release defect density for the Windows Server 2003 sources at a statistically significant level. Zheng et al. show in [ZW+06] that the defect removal yield of static analysis nearly match that of manual inspection. However, software tests have a three times higher yield rate. That means, static analysis cannot replace functional testing but it reaches a good efficiency in improving the code quality and in detecting errors. Moreover, it is a complementary technique to testing that helps the designer to locate bugs as early as possible during development. Ayewah et al. [AP+07] determine that each detected warning does not have a real impact on software functionality. Especially the costs to remove such a warning should be kept low. Therefore, a structured process of defect handling should be established.

2.5.2 General Limitations and Risks

The biggest problem of static analysis arises from the *conservative approximations* made during analysis. It is caused by the undecidability of a static computation of the dynamic runtime behavior. Hence, approximations result in more dependencies than in practise necessary. Concretely, imprecision is caused by a number of language constructs that obviate the precise computation of data dependencies, e.g. the usage of pointers, object orientation, method overloading, or concurrency. Thus, the analysis does not falsely omit any existing dependency but it results in a high rate of reported false warnings. This fact could negatively bias user acceptance, especially in the industrial field.

Example 5. *A typical example that requires a conservative approximation is the access to an array using an index variable, e.g. a[i]. To compute precise dependencies statically, all possible values of 'i' have to be known at the point of access. This could be difficult if 'i' is determined at runtime and depends for instance on user input.*

Besides the imprecision of analysis caused by a conservative approximation, static analysis has a few more flaws:

- *Divergent sources.* The user has to ensure that the statically analyzed source code is always the same that is used later to create the executable. Otherwise, unexplainable and undefined behavior could complicate error search.
- *Abstraction difficulties.* Usually, failure search assumes that the corresponding defect is located only at the source code level. However, a failure could be also caused by the environment, e.g. the compiler, the operating system, or third-party libraries, which should be kept in mind.
- *Analysis power.* Most static analyzers find inconsistent or deviant code instead ensuring the functional correctness of the implemented specification. However, inconsistent or flawed code does not necessarily result in critical misbehavior of the program.

2.5.3 Benefits of REGATTA

The general benefits of static analysis also apply to REGATTA. Additionally, some more advantages exist because of the framework character. In general, these advantages relate to the requirements mentioned on page 41. Especially, FDC specifications have a number of advantages. The underlying FSM provides an easy to learn and familiar concept for most programmers. The strict partition of language-specific settings from the abstract specification supports the creation of a generic code analysis library with a number of advantages:

- *Quick solutions.* For standard analysis problems check implementation speeds up in a significant way.
- *Proven implementation.* A library contains proven solutions of standard analysis problems. Thus, the user has to focus only onto the adaptation to the analyzed language.
- *Subsequent extension.* The abstract level of modeling allows a subsequent extension or change of an FDC specification with little effort.

2.5.4 Limitations of REGATTA

The general limitations and risks of static analysis also apply to REGATTA. In addition, some more limitations exist:

- *Limited application.* A universal and flexible framework solution for all possible static analysis problems is not possible. REGATTA supports only a certain set of use cases (see Section 2.1 on page 41).
- *Learning curve.* If the user wants to implement an analysis, he has to understand the general framework architecture. Furthermore, he has to know the provided framework API. This requires a comprehensive framework documentation and some learning effort.

3 SYSTEMC DESIGN ANALYSIS SYSTEM

Based upon REGATTA, the SystemC static analyzer SDAS was created. It applies the most important framework use case “ensuring coding standards” (see Section 2.1 on page 41). The analyzer implements 18 coding guidelines of a proprietary industrial coding standard. The guidelines range from simple formatting and naming convention rules, applicable to C++ programs, to SystemC specific checks.

Table 3.1: Comparison of REGATTA and SDAS

Criteria	*REGATTA*	*SDAS*
C++ LOC (code + comments)	61,646	8,577
... generic components	61,304	1,117
... (pre-defined) guideline checks	342	7,460
LOC of LALR(1) SystemC grammar (code + comment)		
... scanner specification	-	570
... parser specification	-	3,128
lines of FDC XML configuration		
... symbol table	-	189
... DFA	-	124

3.1 Implementation Effort

The analyzed code is parsed by a yacc compatible C++ LALR(1) grammar from Willink [Wil01]. This grammar was extended to recognize SystemC constructs. The lex/yacc front-end generator of REGATTA creates the scanner and the parser. A direct recognition of SystemC language constructs enables the easy check of SystemC specific coding guidelines such as the report of obsolete SystemC 2.0 constructs.

At the back-end side the generic symbol table and the DFA components were configured using the provided FDC XML configuration interface. The particular coding guideline checks were implemented using the provided framework API.

Table 3.1 compares various characteristics concerning the implementation effort of the REGATTA framework and the SystemC analyzer SDAS. Using the framework, 1117 lines of C++ code (LOC), i.e. only 1.82% of the amount of framework code, are required to create a working back-end for SystemC analysis. In addition, there is a number of approximately 3,700 lines of grammar description to specify the front-end. The major amount of code in SDAS (7,460 lines) represents guideline checks. Remarkable is the small number of 213 XML lines to configure the generic symbol table and the DFA component to SystemC. It shows that only a few lines of configuration code are required to adapt generic framework functionality to a concrete language.

3.2 Configuration and Adaption

The generic framework components (see Section 2.3 on page 43) are adapted using the FDC XML configuration interface. The FDC specifications are tailored to SystemC using the analysis alphabet A_{SC} defined over the given SystemC grammar.

```
<Rule>
  <Id>T-3</Id>
  <Switch>ON</Switch>
  <ImplBy>AU_FDC_ComponentCnt</ImplBy>
  <Severity>2</Severity>
  <Description>Define only one SystemC module/C++ class per header file</Description>
  <FDCCfg>
    <Import name="max_cnt">1</Import>
    <Import name="pat">.*\.h$</Import>
    <PKey name="PK_CNT">RK_SC_MODULE_DECLARATION_1_0|RK_CLASS_SPECIFIER_1_0</PKey>
    <PKey name="PK_INIT">RK_LOCALPRE</PKey>
    <PKey name="PK_RESET">RK_LOCALPOST</PKey>
  </FDCCfg>
</Rule>
```

Figure 3.14: Configuration of the FDC "ComponentCnt"

Example 6. *Figure 3.14 shows the configuration of the abstract FDC "ComponentCnt" from Figure 3.13. The concrete check shall verify that only one class or SystemC module definition is defined per header file in an analyzed SystemC design. To do that, the user has to specify the imported variables, i.e. sets the count limit to '1' and the filename pattern to '.*\.h$'. Moreover, the proxy symbol PK_CNT is assigned two symbols from A_{SC} which introduce the definition of a new class or SystemC module. The symbols PK_LOCALPRE and PK_LOCALPOST are implicitly given by REGATTA. These symbols indicate start and end of the analysis of a new file, i.e. are used to initialize and to reset the check. The other XML tags specify particular information that are output in case of a found coding violation, e.g. the error description.*

Similar to Figure 3.14, the generic symbol table and the DFA component are tailored to SystemC. The following examples illustrate some analysis results of SDAS.

Example 7. *Figure 3.15(a) shows a small C++ function which calculates the sum of the first "num" naturals. Figure 3.15(b) depicts a dump of the corresponding SDAS symbol table after analyzing the function. The dump describes the function "sum", at first. Within this function the local integer variable "s" is declared. Next, an unnamed block opens the scope of the for-loop body.*

Example 8. *Figure 3.16 illustrates the CFG with annotated variable accesses for the "sum" function presented in Figure 3.15(a). The annotated CFG is created by the tailored DFA component. Since REGATTA currently does not support an interprocedural DFA, nothing can be assumed about the value of the function argument. So, the function argument "num" in line 1 is only labeled as declared, i.e. '(D)'.*

```
1 void sum(int num)
2 {
3   int s = 0;
4   for (int i = 1; i <= num; ++i)
5   {
6     s += i;
7   }
8   std::cout << "sum of first " << num
              << " naturals = " << s << std::endl;
9 }
```

(a) Calculate the sum of the first "num" naturals

```
### SYMBOL TABLE DUMP ###
  sum (Method)
   Location: 1,5 in file deduction/simple_ex.cpp
   Dispatched: 0
   Result Type: void
   Parameter: num (int)
   Qualification: No qualification
     sum.s (Variable)
      Location: 3,6 in file deduction/simple_ex.cpp
      Type: int
        sum.s.int (Primitive Type)
         Location: -1,-1 in file implicit
      Owner: type of sum
     sum.unnamed_blk_0 (Grouped Type)
      Location: 4,31 in file deduction/simple_ex.cpp
```

(b) Symbol table dump of the "sum" function

Figure 3.15: Symbol analysis in SystemC

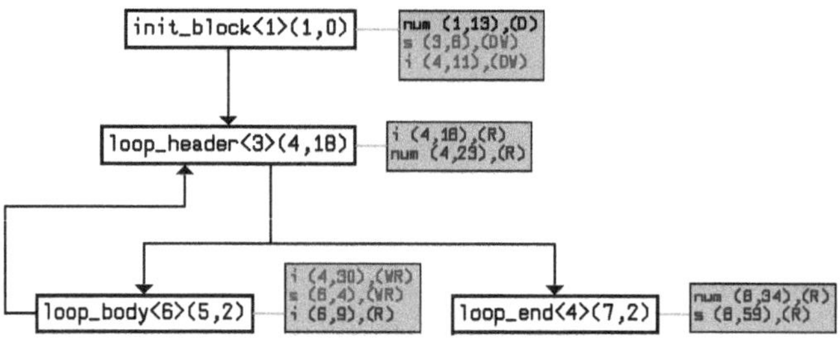

Figure 3.16: CFG with annotated variable accesses

Example 9. *The check for undefined variables (see Figure 3.10) on the "sum" function from Figure 3.15(a) reports two violations. They are caused by reading the variable "num" at rows 4 and 8. These two errors are found due to the intraprocedural DFA which cannot determine whether the function argument is assigned a valid value at the call side of the "sum" function. Figure 3.17(a) shows the* REGATTA *analysis report while Figure 3.17(b) depicts the annotated CFG.*

Example 10. *The consistent usage of design patterns in a SystemC design is a good indicator for more reusable, maintainable, and better readable code. A pattern which can be structurally detected, is the Composite design pattern [GHJV95]. In Figure 3.11 the proper CrocoPat query is already sketched. Figure 3.18 presents an example of three created relations after parsing the given SystemC design.*

```
1 file(s) to analyse
analysing file deduction/simple_ex.cpp
3-1 Variable >num< undef in basicblock 3 on path:
    3 ->1 deduction/simple_ex.cpp 4 23 4 26
3-1 Variable >num< undef in basicblock 4 on path:
    4 ->3 ->1 deduction/simple_ex.cpp 8 34 8 37
```

(a) Analysis report

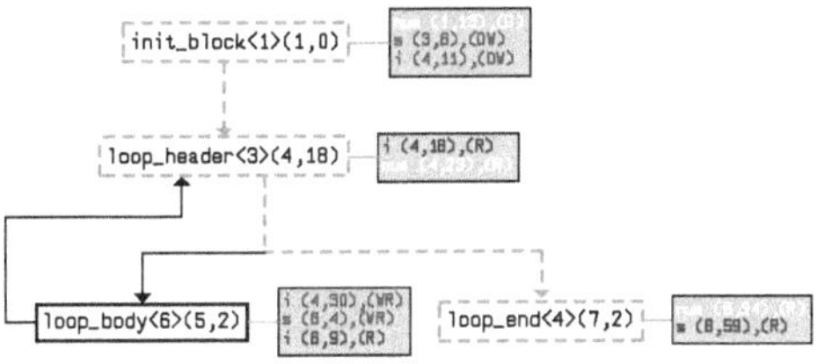

(b) Annotated CFG with violated paths

Figure 3.17: Detecting undefined variable accesses

```
class top : public sc_module
{ ... };
class contained
{ ... };
class child : public top
{
  contained attr;
};
```

(a) SystemC example design

```
INHERIT top sc_module
INHERIT child top
CONTAIN child contained
```

(b) Created relations for the example

Figure 3.18: Created relations in a SystemC design as used by *CrocoPat*

3.3 Example Analysis Flow

Figure 3.19 sketches an exemplary analysis flow while running a concrete SystemC analysis. Generally, the flow divides into three main phases: process the input, perform activated analyses, and output the obtained results:

1. *Process input.* The front-end parses the input sources consisting of a single file or a collection of files. As defined in Definition 4, an EST is created by the front-end in case a valid derivation could be found for the input.
2. *Perform analysis.* Before static analysis is started, each analysis task registers at the front-end for the actual needed data. There, the analysis alphabet A_{SC} defined over the given SystemC grammar is used. The grey colored tree nodes in Figure 3.19 shall sketch registered data. The parse points in the analysis alphabet have a special meaning. They are used to control the analysis process and allow to collect context-sensitive information. Figure 3.20 presents an example. Here, the parse point RK_SC_METHOD_1_2 denotes that the following meaningful token contains the name of an SC_METHOD process declaration and initiates its collection.

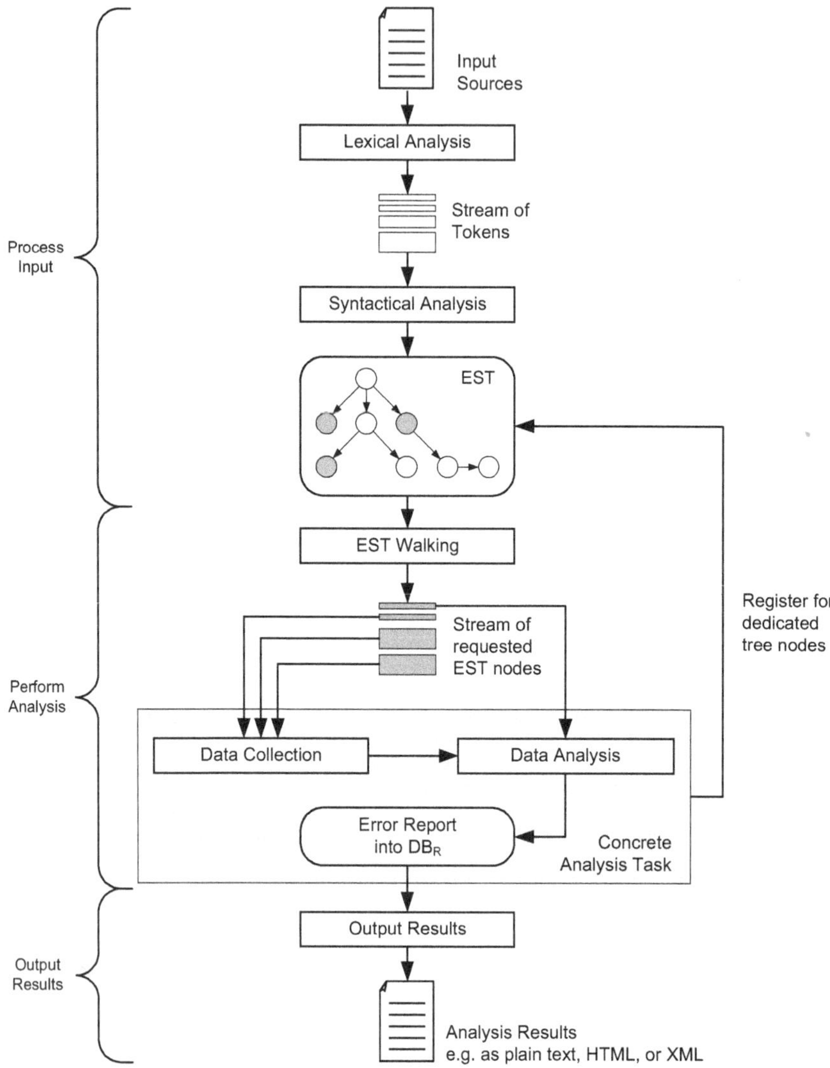

Figure 3.19: REGATTA exemplary analysis flow

After the input is completely parsed, the EST is walked in a depth-first search. Thereby, only registered nodes are forwarded for analyses. If all required data are available, they are checked. In consequence of a coding violation, a detailed error report is stored into the report database DB_R. A report consists of position information (row, column, filename), a unique error code, and a user-defined field. This field can

Grammar rule for SC_METHOD declarations annotated with parse points

```
sc_method:
RK_SC_METHOD_1_0 SC_METHOD
RK_SC_METHOD_1_1 '('
RK_SC_METHOD_1_2 id
RK_SC_METHOD_1_3 ')'
RK_SC_METHOD_1_4 ';'
RK_SC_METHOD_1
```

Procedure to collect all SC_METHOD process names:

1. Register for parse point `RK_SC_METHOD_1_2`.
2. The node representing the requested parse point is submitted to the back-end.
3. Use framework method `FE_Data::getNextLRT()` to retrieve the next meaningful token, i.e. the process name.

Figure 3.20: Collect names of found `SC_METHOD` processes

hold additional information to describe the detected violation in more detail.

3. *Output results.* After finishing all analyses, the results are read from the DB_R and written into various output formats. REGATTA supports amongst others plain text, XML, and a specific report format for the XEmacs editor.

4 EXPERIMENTAL RESULTS

The following section considers two experiments. The first one demonstrates the strengths of dataflow anomaly analyses while detecting a serious functional error in the SIMD data transfer example (see page 28). The second experiment applies SDAS to the particular development stages of an industrial SystemC-based verification environment over a period of four years.

4.1 SIMD Data Transfer Example Continued

According to Figure 3.1, static analysis is applicable as soon as first syntactical correct blocks of code have been written. Now, we assume that the designer has coded the processor unit as one of the first SystemC modules of the SIMD data transfer example. For test purposes, the address to access data in the memory is calculated at random using the local function `proc_unit::get_addr()`. The random address value is assigned to the private class attribute `proc_unit::addr`. Then, this address is transmitted as generic payload attribute of a read or write transaction. Figure 3.21 shows a code snippet which creates a correct write transaction. This transaction writes data into the memory.

Now, we want to assume that the designer has forgotten to request a new address, i.e. line 94 has not been coded. Provided that the read transaction receives a correct address, the processor unit would always write data to the address the last data were read from memory because the variable `addr` is a

```
..
84 // ----------------------------------------------
85 void proc_unit::write_data()
86 {
..
94    addr = get_addr();
95
96    // set write transaction parameters
97    tlm::tlm_command cmd =
98     static_cast<tlm::tlm_command>(tlm::TLM_WRITE_COMMAND);
99    w_trans.set_command(cmd);
100   w_trans.set_address(addr);
...
```

Figure 3.21: Create a write transaction in the SIMD processor unit

```
Analysis started at 15:59:52 on 3.9.2009
   2 file(s) to analyse
   analysing file r2c_undef/proc_unit.h
   analysing file r2c_undef/proc_unit.cpp
   ...
   3-1 Variable >addr< undef in basic block 23 on path:
      23 ->21 ->20 r2c_undef/proc_unit.cpp 100 24 100 28
   3-1 Variable >addr< undef in basic block 23 on path:
      23 ->21 ->20 r2c_undef/proc_unit.cpp 109 50 109 54

  Analysis finished at 15:59:52 on 3.9.2009
```

Figure 3.22: SDAS error report indicating an undefined address in case of write transactions

shared class attribute. Obviously, this could produce a serious misbehavior if the model is simulated during later development stages.

SDAS reports this coding error by performing a dataflow anomaly analysis that points to potentially undefined variable accesses. Figure 3.22 documents the analysis outcome which had identified two problematic read accesses. The first read access sets the address as generic payload attribute (line 100 in Figure 3.21). The second one corresponds to a debug output of the address value in line 109.[1] The error message supports debugging by reporting the possible execution path through the program on which the variable is undefined.

1. This line is left out in Figure 3.21 due to space limitations.

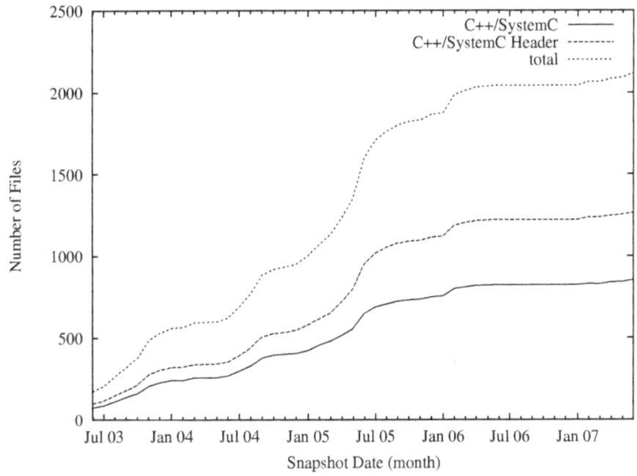

Figure 3.23: Number of analyzed files

4.2 Industrial SystemC Verification Environment

SDAS was used to evaluate the quality of an industrial SystemC-based verification environment in terms of a set of 18 company-specific coding guidelines. The verification environment consists of a number of core components encapsulating base verification functionality and several concrete test benches. It has been developed to verify hardware designs using a co-simulation between the RTL design and the SystemC test bench. A proprietary TLM approach together with monitor components, BFM, transaction generators, reference models, and check transaction dispatchers define a general verification infrastructure. This infrastructure facilitates the efficient development of SystemC-based test benches.

The coding guideline set was published in August 2003. A stable version of SDAS was available in November 2003. The usage of SDAS was integrated into the development flow on a voluntary basis. For experimental evaluation CVS snapshots of the SystemC verification environment were analyzed. Each snapshot was extracted at the first of the month over a period of four years starting in June 2003 until June 2007. This period embraces a major part of the development time of the verification environment. As can be seen in Figures 3.23 and 3.24, much code was written starting from 15,781 lines of code (including comments) distributed over 174 files in June 2003 to 316,709 lines of code and 2,118 files in June 2007.

Figures 3.25 and 3.26 document the number of coding violations found in each monthly CVS snapshot. Figure 3.25 selects typical representative coding guidelines each taken from another error severity class, namely *critical*, *major*, *minor*, *warning*, and *information*. The error severity helps the user to plan bug fixing according to the significance of a violation. As can be

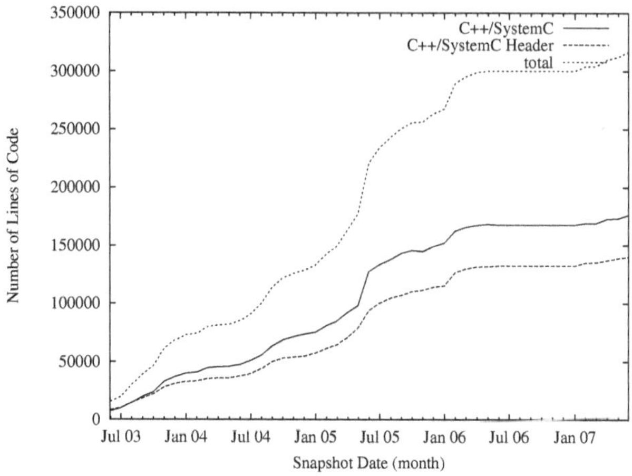

Figure 3.24: Number of analyzed lines of code

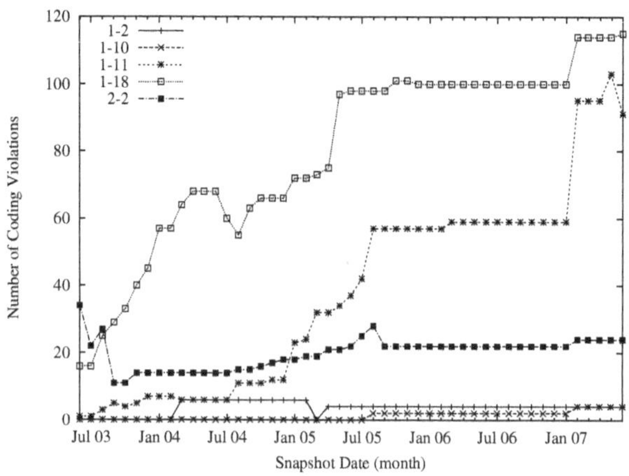

Figure 3.25: Number of violations per month for selected coding checks

seen, the number of detected violations constantly increase over time except for guideline 1-2.[2] Remarkably, there exist some reductions of the number of violations in case of specific guidelines, e.g. guideline 1-18[3] between May 2004 (68) and July 2004 (55), or guideline 2-2[4] between June 2003 (34) and September 2003 (11). A review of the CVS logs revealed manual code reviews as cause of reduction. In this reviews crucial source code part were

2. Guideline 1-2. *Name file like containing class declaration.*

3. Guideline 1-18. *Reconvergent inheritance is not allowed.*

4. Guideline 2-2. *Do not use obsolete SystemC 2.0 constructs.*

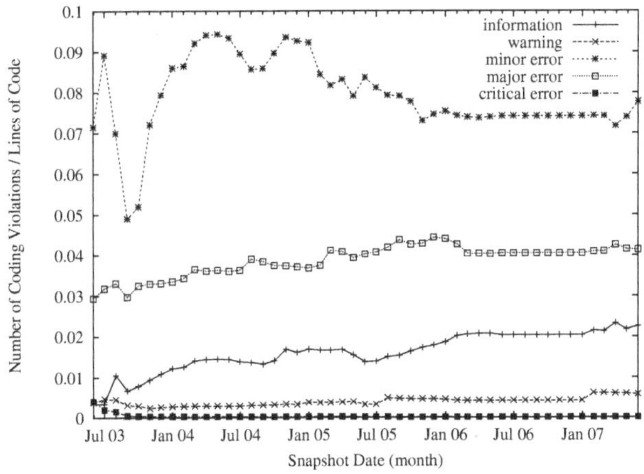

Figure 3.26: Violations per line of code for all coding checks

identified for bug fixing. Other project priorities mostly prevented a more thorough fixing. A comparable picture gives Figure 3.26. It presents the number of violations per code line by summarizing guidelines by severity classes. Except for the critical severity, the number of violations in all other severity classes rises constantly over development time. Critical coding violations decrease because only coding guideline 2-2 is classified as critical and the reported violations are small and hardly increase during development (see Figure 3.25).

The verification environment was under development over the considered period of time. Thus, it was subject of different architectural changes and refactoring steps, e.g. renaming or deletions of files, functions, or methods. So, the aforementioned reduction in the number of coding violations could be also caused by a simple deletion of files which would not represent a real fix. Figure 3.27 shows all violations found in the 166 files that exist over the whole considered period of time. These files mainly describe core functionality of the verification environment. As can be seen in the figure, only three coding checks, i.e. guideline 1-11,[5] 1-18, and 2-2, report violations. During the first months many of them have been fixed. This was the time the general architecture of the environment was specified and implemented. After this phase, fixing occurred only sporadically.

Summarized, more coding violation than primarily expected were found. SDAS detects 31,349 coding violations in 31 s over 316,709 lines of code on the June 2007 snapshot.[6] There, the majority of violations (25,214) is

5. Guideline 1-11: *Each typedef should reference a distinct type.*

6. Test system: OS Linux, CPU 2.4 GHz AMD Opteron™ 180, 4 GB RAM.

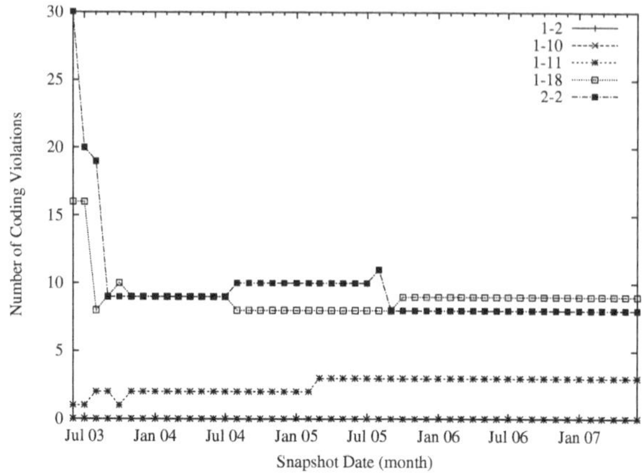

Figure 3.27: Number of coding violations existing from the beginning

caused only by three minor important guidelines which primarily ensure code formatting. Even though most violations are easy to fix, the developers often spend little time for fixing. The reasons are complex such as moved project priorities, or too narrow deadlines. Of great importance is also the precision of analysis results. If the developer is faced with a, personally felt, too high number of false positives, he will reject any further tool usage very fast.

5 SUMMARY AND FUTURE WORK

Improving the verification efficiency by static code analysis has been shown in this chapter. Static analysis finds functional errors and ensures the code quality of SystemC modules long before the system model gets simulatable the first time. The introduced analysis framework REGATTA assists the generation of tools for static analysis. FDC specifications provide an easy to use and powerful configuration approach that facilitates the implementation of analyses and generic framework components. There is no other analysis framework known that has been used for so many languages and analyses, e.g. to ensure coding standards for Verilog, *e* [Rog02], SystemC, to generate library documentation for VHDL-AMS models [RFSH05], or to translate Verilog designs [HR08].

The usability of static analysis techniques is demonstrated by using the SystemC analyzer SDAS. SDAS has been applied onto the SIMD data transfer example. Hence, functional errors in (parts of) a system model can be detected without any need for simulation. Moreover, SDAS implements a proprietary industry SystemC coding standard whose compliance has been checked on a complex real-world SystemC verification environment.

Summarized, static analysis makes important contributions to facilitate debugging. In general, related tools scale very well to large design descriptions, i.e. several million lines of code. For instance, SDAS analyzes over 300,000 lines of code in several seconds. The found coding violations simplify manual code reviews, help the developer to follow a corporate coding standard, and point to typical pitfalls and coding flaws. The efficiency of static analysis has been documented by several empirical studies. It is nearly as effective as manual inspection [ZW+06] while 60% of defects in software programs can be caught by peer reviews [Shu02]. A study from Bloor Research [PC95] comes to similar results where 60% of software defects found in released software products could have been detected by means of static analysis. So, static analysis is an efficient debugging technique. However, its successful application needs an integration into the development flow. This comprises fixed timeslices in the project schedule as well as coding standards that will be supported by the majority of the developers.

Possible future work could embrace the development of an FDC-based generic code analysis library that provides proven implementations for many standard analysis problems.

Chapter 4

High-Level Debugging and Exploration

Up to now, the designer has checked code quality and particular functional correctness aspects of the system model using static analysis. Now, the first executable version of the model can be compiled and simulated. This version may only comprise a subsystem of the final system. In place of static tests dynamic analysis in terms of *observation techniques* support the designer in debugging the simulatable design description. If the simulation produces an erroneous outcome the simulation state is observed at interesting moments in time. Therefore, a debugging and exploration approach at a higher abstraction level is proposed (see Figure 4.1).

Generally, there are several observation techniques that allow to examine program runs in order to find program errors. Logging, a powerful but simple technique, is supported by many programming and description languages. Here, the designer manually inserts proper logging statements into the source code. Then, these statements write interesting program facts to some output devices.

The second observation technique uses a debugger. This external tool observes the program state without the need for any code modifications. A debugger hooks into the code and accompanies the program execution while dynamically monitoring user-defined, arbitrary program facts.

In case of complex programs a usual textual representation of the program state could be insufficient. Especially, the architecture of the program and the relationship between different components and subsystems cannot be easily surveyed. Hence, advanced visualization techniques improve system exploration and accelerate the debugging process.

The first part of the chapter introduces state-of-the-art observation techniques used for arbitrary programming languages in general. Furthermore, their support in the SystemC context is discussed. On that basis, various requirements for an integrated system-level debugging environment are defined. Next, a debug flow is presented that guides debugging and exploration of system models.

The second part of the chapter describes SHIELD (Systematic HIgh-levEL SystemC Debugging) – an integrated debugging environment for SystemC. So called debug patterns provide particular debug strategies. A strategy, by means of a debug pattern, presents a formalized procedure to fix a bug that is notified by an always recurring failure symptom. Using a non-intrusive

F. Rogin and R. Drechsler, *Debugging at the Electronic System Level*,
DOI 10.1007/978-90-481-9255-7_4, © Springer Science+Business Media B.V. 2010

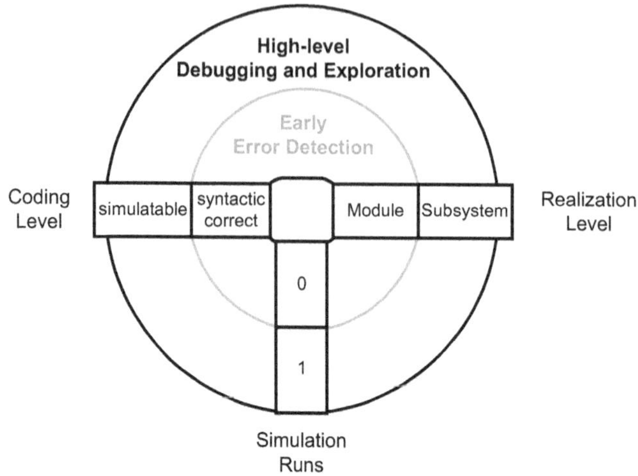

Figure 4.1: Debugging and exploration of system models at a higher level

SystemC debugger enables the developer to debug arbitrary, unaltered designs. This comprises debugging not only at the algorithmic, untimed level but also at the system level, e.g. to handle loosely-timed, or approximately-timed design descriptions. The debugger explicitly supports SystemC concepts and the SystemC simulation semantics. As result, the designer gets quick and concise insight into the static structure and the dynamic behavior of the design without the burden of gaining a detailed knowledge of the underlying SystemC simulation kernel. Visualization and exploration techniques facilitate the debugging process of complex designs.

1 OBSERVATION TECHNIQUES IN A NUTSHELL

First, this section gives a short overview about general observation techniques and their specific SystemC support. A detailed summary of observation techniques is given for instance by Zeller [Zel05]. Finally, the section closes with a discussion of related work.

1.1 Overview

In contrast to deduction techniques such as static analysis (see Chapter 3), observation is a dynamic technique that explores concrete program runs. In case of a failing run, it is observed what has been actually happened. Subsequently, the user tries to determine what has caused the failure. This process of cause analysis is also called *diagnosis*.

1.1.1 Logging

Logging is the simplest and most widespread observation technique. This technique provides a set of (specialized) logging statements that are manually inserted into the program code. So, logging statements write user-defined facts about the concrete program state to some output device such as the text console or a log file. The drawbacks of this technique are for instance cluttered code, huge log reports mixed with the ordinary output, or a performance loss. To encounter these drawbacks logging should follow several rules. This includes the output in standard formats that ease filtering and searching of log information. An optional on/off switch can avoid effects on the performance. Moreover, a variable granularity supports the designer to focus logging on dedicated components or verbosity levels.

Logging facilities are already supported by SystemC built-in trace and report mechanisms, i.e. `sc_trace` and `sc_report` statements. Here, the designer manually inserts logging statements into the particular SystemC modules.

1.1.2 Debugging

Debuggers are well-known and widespread tools that hook into the execution of a program without the need for any code modifications. In contrast to logging, an interactive debugger can dynamically observe arbitrary facts of the inspected program. Nearly each debugger provides the following main features: execute and stop the debugged program (on specified conditions) at any program point, observe and change the program state. In addition to an interactive debugger, a postmortem debugger reads in memory dump files that were created in case a program had crashed. The backtrace of such a dump gives valuable hints what was the system state at the time of the crash. One of the most powerful debuggers is the GNU debugger GDB [GDB]. GDB supports many different languages such as C, C++, Pascal, Fortran, or Ada.

Debugging can be supported by *program slicing* that was originally introduced by Weiser [Wei84]. Often during debugging a variable v at some execution position q holds an incorrect value. So, it comes in handy to know which parts of the program are relevant to search for the error. The program slice with respect to v at execution position q encompasses all statements of the program that might affect the value of v. Precisely speaking, this is the definition of a *backward slice* $S_s^B(v, q)$ where (v, q) is called the *slicing criterion.* In contrast, a *forward slice* $S_s^F(v, q)$ encompasses all statements of the program that could ever be affected with respect to v at execution position q. Korel and Laski [KL88] introduced *dynamic program slicing* – the dynamic counterpart of static slicing. There, only those statements are part of a slice that *really* affected the value of a variable at some statement. Hence, a

```
1  p = 0;
2  i = 1;
3  c = 1;
4  n = read();
5  g = read();
6  if (g = 1)
7     c = 10;
8  while (i <= n) {
9     p = p + i;
10    if (c > 1)
11       p = p * c;
12    i = i + 1;
13 }
14 write(p);
```

(a) Static slice for *p* at line 14

```
1  p = 0;
2  i = 1;
3  c = 1;
4  n = read(); // 3
5  g = read(); // 0
6  if (g = 1)
7     c = 10;
8  while (i <= n) {
9     p = p + i;
10    if (c > 1)
11       p = p * c;
12    i = i + 1;
13 }
14 write(p);
```

(b) Dynamic slice for *p* at line 14

Figure 4.2: Static vs. dynamic slicing

dynamic slice is small as compared to the respective static slice. Consequently, a dynamic slice is valid only for a given program input x, i.e. a concrete program run. So, the slicing criterion for the dynamic backward slice $S_d^B(v, q, x)$ and the forward slice $S_d^F(v, q, x)$ is (v, q, x).

Example 11. *Figure 4.2 contrasts static and dynamic slicing in case of variable p being output at line 14. The static backward slice* $S_s^B(p, 14)$ *comprises the entire program. The dynamic backward slice applies to a program run that sets* n=3 *and* g=0, *i.e.* $S_d^B(p, 14, \{3, 0\})$. *Here, over 33% of all statements do not influence the value of p. So, these statements do not need to be considered during debugging.*

Currently, SystemC does not comprise sophisticated debugging features. Rather, standard C++ debuggers are usually used to analyze the simulation state during an erroneous simulation run. Unfortunately, a C++ debugger operates on a very low abstraction level, usually the algorithmic level (see Figure 2.2). It does not understand specific SystemC constructs nor their semantics. Another point is that with the parallel development of SW and HW, also design sizes and complexities tend to increase. Thus, it becomes less obvious where to start and which blocks to observe in a debugging process. Moreover, language features such as multi-threading and event-based communication increase the program complexity and introduce nondeterminism in the system behavior. Consequently, debugging SystemC designs is very challenging and should be specially supported.

1.1.3 Visualization

Classical debugging approaches often rely on a textual output of the program state at particular points in time. Visualization could help the developer to keep track of complex data structures, the architecture of the program, or the relationships between different components. Some debuggers provide a graphical front-end such as the GNU Data Display Debugger for GDB. In hardware design, visualization assists the designer in getting insight into the structural design hierarchy, or the connectivity of components. Currently, SystemC does not integrate any direct visualization support.

1.2 Related Work

Observing a concrete SystemC simulation run requires hybrid techniques that grant a quick access to design components but also allow to evaluate ordinary C++ code. Unfortunately, C++ fragments cannot be reached by using SystemC data introspection techniques. Even though there are several commercial and academical tools supporting SystemC debugging, only few of them offer an advanced visual interface to the designer.

RealView Debugger Suite [ARM] comprises a complete integrated development suite that allows to implement, to simulate, to debug, and to analyze SystemC/C++ designs. It addresses architectural analysis as well as SystemC component debugging at algorithmic up to the transactional level. Especially the debugging of embedded applications (running on remote targets such as ARM processors) is supported. The *Platform Architect* development environment [CoWare] targets system level design and verification based on the Eclipse framework. It utilizes a native simulation environment which is specially adopted to fit SystemC needs. The integrated debugger offers specific commands supporting source and system level breakpoints as well as thread debugging. Additionally, the user can initiate a graphical tracing of SystemC events, threads, and transactions. Contrary to our debugging environment SHIELD, both commercial solutions come with their own vendor-specific SystemC kernel. This fact prevents the easy integration into an already existing design flow. Moreover, debugging is not supported by a debug flow as implemented by SHIELD that guides a systematic debug procedure.

The GRACE++ system [WD+05] uses SystemC simulation results to create Message Sequence Charts in order to visualize and analyze inter-process communication. Various filters help to reduce information complexity. The approach presented in [CRAB01] applies the observer pattern [GHJV95] to connect the external evaluation software to the SystemC simulation kernel. This general method facilitates loose coupling but requires possibly undesired modifications of the kernel.

One of the first approaches that accomplishes SystemC design visualization has been introduced in [GDLA03]. The implementation uses the SystemC kernel to analyze models during execution. An interactive graphical back-end facilitates the design visualization. Even though models can be specified using C++ features, analysis and visualization are limited to SystemC objects. There, only the data flow can be viewed where behavioral information is not available. Since this approach has to execute the model without further information about declarations, it is not aware of detailed positional information regarding the instantiated objects. Hence, crossprobing facilities are very restricted.

Another approach that facilitates designers in visualizing SystemC models is presented in [EAH05]. Since it is based on data introspection too, it shares many restrictions with [GDLA03]. One major difference to [GDLA03] is the usage of a graphical user interface that has been especially designed for this approach. However, the visualization interface does not support features like crossprobing for path fragment navigation. Contrary to the aforementioned work, SystemCXML [BP+05] and LusSy [MMM05a] do not use data introspection for the purpose of analysis. While the extraction of the hierarchy in SystemCXML is done via Doxygen, LusSy uses PINAPA [MMM05b]. The visualization is realized as graph structures generating control or data flow graphs.

Several work propose methodologies that are similar to our concept of debug patterns. The debugging environment MAD [KSF99] is based on event graphs that are constructed from recorded event traces of parallel program runs. A subsequent graph analysis detects errors and anomalies automatically. MAD allows to specify communication patterns which define the expected behavior of the program. These patterns are checked against the event graph and the results help the user to focus his attention on the most critical parts in the graph. A quite similar approach of pattern-oriented debugging was proposed by the TAU programming analysis environment [SC+96] which uses the event-based debugger Ariadne [CF+93]. Ariadne matches user-specified models of intended program behavior against the actual program behavior captured in event traces.

MAD and TAU focus on the interprocess communication of massively parallel programs operating on monitored traces. There, the entire debugging process is done in a post-processing step. In contrast, SHIELD does not monitor and evaluate traces but provides debugging support directly at runtime especially tailored to SystemC needs. While both approaches rely upon proprietary debug tools, our solution is implemented on top of the GNU debugger GDB.

An early work [DE88] underlines the importance of having knowledge of likely defects to ease debugging. The authors introduce the notion of a *stereotyped bug* which describes the fact that the same couple "symptom(s) + bug" has appeared several times. A symptom can be an I/O discrepancy, a trace content, or a specific program state. SHIELD uses similar symptoms to propose a proper debug pattern. In [DE88] different tools are mentioned that perform a recognition of stereotyped bugs by defining appropriate debug procedures. Most of these tools are aimed at logic programming systems where the applied techniques cannot be adopted directly to an imperative language like SystemC.

The authors in [KP99] present some loose hints and generally accepted approaches to find defects faster. Examples are the *divide-and-conquer* approach to isolate the point of failure or the evaluation of logged trace data. Instead, our debug patterns are presented in a formalized way especially aiming at the SystemC debugging needs. The debugging environment guides the user through the debug process and supports him by partially suggesting applicable patterns.

Summarized, none of the listed tools and approaches work with the OSCI SystemC reference kernel implementation, an unpatched GDB, support a high-level debugging interface, and additionally offer a sophisticated visualization of SystemC designs. Hence, the mentioned tools and approaches do not fit seamlessly into an existing design flow. Furthermore, a systematic debug flow as provided in SHIELD is missing.

2 SYSTEM-LEVEL DEBUGGING

First, this section itemizes several requirements that should be followed to develop a debugging environment for ESL designs. Second, a methodology is proposed that results in a systematic error search at the system level.

2.1 Requirements

The development of a debugging solution for the system level is driven by user demands, the conditions of use, and the features expected from a debugger.

- *Non-intrusiveness*. The solution should work with an unmodified version of the used SDL. It shall avoid any changes to present designs or (third-party) IP blocks. Such a non-intrusive approach fits seamlessly into any existing design flow which reduces maintenance and customization. On the tool side, a powerful and popular debugger such as

GDB can form a solid development basis which should be extended without any need for patching its sources. Advantages are an intuitive and unchanged debugging flow combined with a minimal learning curve for users who are already familiar with particular debugger tools.

- *System-level debugging information.* Information at system level should be retrievable fast and easily. According to [DSG03], three information categories are of interest: (i) *Static simulation information* describe the structure of the system architecture such as existing modules, subsystems, processing units, signals, I/O interfaces, or the communication architecture. (ii) *Dynamic simulation information* include amongst others the trigger conditions and the sensitivity lists of processing units, available synchronization events, or the values of signals logged over a period of time. (iii) *Debugging callbacks* allow to add callbacks from the simulation environment to break simulation on certain events such as process activations, value changes on signals, or the ongoing simulation time. To support debugging of complex system designs sophisticated visualization and exploration features should be supplied, as well.
- *Debugging features.* The debugging solution should provide specific system-level debugging commands that directly build up on the syntax and the semantic of the used SDL. Thus, debugging commands become language-aware. Different command classes shall facilitate an efficient debugging. (i) *Examination commands* retrieve either static or dynamic simulation information. So, in case of a failure the user gets a fast insight into the relevant design parts and their relationship. There, a number of commands allow to interactively control the visualization of the design and its simulation state. (ii) *Controlling commands* provide stop and step functionality at system level that means to halt and continue the simulation at specific conditions.

2.2 Methodology

This section details the proposed debugging methodology. The available debug levels are introduced at first. These levels form the basis for a strategy that is presented at next. Finally, dynamic program slicing for simplified debugging is discussed.

2.2.1 Debug Levels

Three debug levels are the basis for the proposed debug flow. These levels are supported by dedicated methods to locate and correct defects in a system

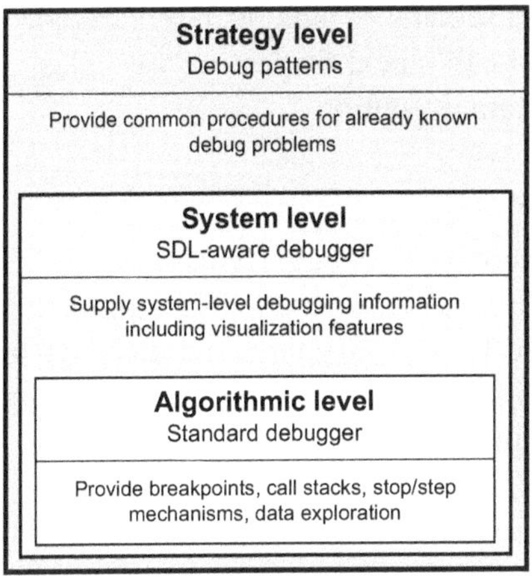

Figure 4.3: Hierarchy of debug levels

design. Figure 4.3 shows the hierarchy of debug levels. Debug levels should not be mixed up with abstraction levels introduced for ESL design (see Section 1.1 on page 10).

- *Strategy level.* At the strategy level, debug patterns define a strategy to guide the overall debugging process. In case of the occurrence of a familiar failure symptom the user gets some help in form of a guidance how to locate and fix the failure-causing defect. Such a failure symptom could be for instance an infinitely looping simulation or the output of unexpected simulation values. A particular pattern is based on the debugging functionality provided at system level.
- *System level.* At system level, the debugging environment enables the user to debug a design at the ESL. This level is additionally supported by a visualization of the design components. Here, the architecture and/or the interaction between design parts are responsible for failures such as an erroneous communication or the faulty integration of an IP block. The environment supplies system-level information including static and dynamic simulation information. As result, it offers various sophisticated SDL-aware debugging features. If the error turns out to lie at the algorithmic level, this level will be entered and standard debugging commands are used.
- *Algorithmic level.* At the algorithmic level, standard debugging features are applied. Hence, the algorithmic implementation is analyzed during a concrete simulation run such as algorithms or data structures. Standard debuggers supply different capabilities to investigate low-

level program details which usually include setting breakpoints, examining the call stack, inspecting variables, and stepping or stopping the program execution.

2.2.2 Debug Flow

Based on the hierarchy of debug levels, a three-level strategy for error detection in system designs is proposed. This approach leads to a debug flow as presented in Figure 4.4 which guides the designer to a systematic debug procedure. At the beginning, there is always a flawed simulation which is represented by several failure symptoms such as

- a crashed simulation,
- an abnormal long running simulation,
- a waveform mismatch in case of a co-simulation with a hardware design,
- an unexpected debug message, or
- a self-detected error through generated monitors or assertions.

As proposed in [DE88], these failure symptoms could be used to initially guide the debugging process. After a detailed manual analysis of the occurred symptom, the designer looks for a suitable debug strategy. If possible, the debugging environment partially suggests matching debug patterns or automates some actions in order to find the probable cause of an observed symptom more quickly. Otherwise, the designer manually chooses a particular debug pattern. The pattern describes a procedure by means of a sequence of system-level debugging commands to locate the defect initially causing the failure symptom. For an exact detection of algorithmic misbehavior, standard debugger commands are applied subsequently. If the defect is successfully identified, the design is fixed and the simulation is restarted. Otherwise another suitable debug pattern is chosen. If the observed failure symptom does not match any debug pattern, the designer continues working directly at system and algorithmic level, respectively. At these levels, visualization features allow a fast and easy exploration of the debugged system design that help to focus the designer on relevant design parts.

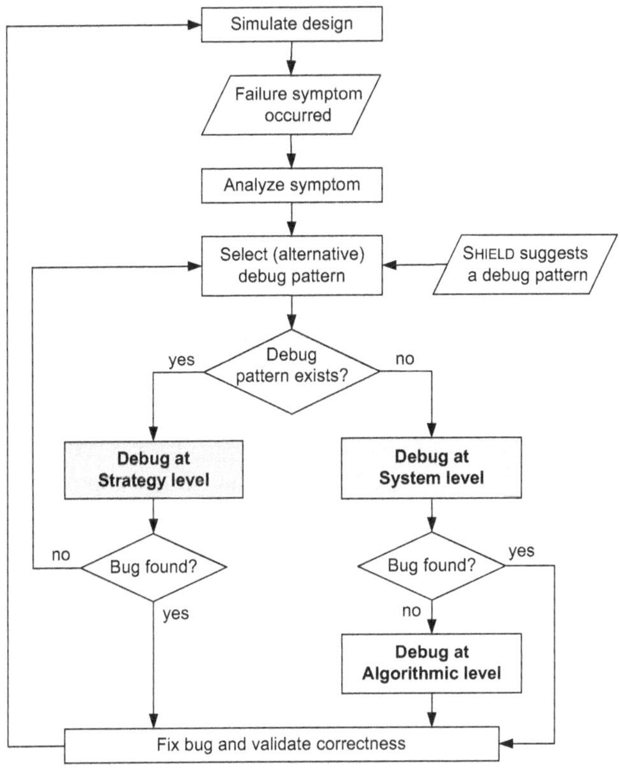

Figure 4.4: Debug flow to guide debugging at the ESL

The utilization of a systematic debug flow provides comprehensive debugging support. Hence, the user can focus his attention to a higher design level (while ignoring irrelevant design parts) through

- a systematic process suggesting applicable debug patterns,
- a formalized procedure to fix already known defects, and
- an improved usability becoming faster familiar with the provided debugging features.

The introduction of such a flow simplifies and accelerates debugging of system designs. Especially the novice user can exclude and fix defects using debug patterns before he has to consult an expert. A further advantage is the intuitive partially automated flow resulting in a defined proceeding. The designer can focus on the underlying failure cause instead of concentrating on the correct debugger usage. A disadvantage of debug patterns is that only patterns for always recurring failure symptoms most probably caused by the same defect or class of defects are available. Consequently each problem, in particular the rare and tricky failures, cannot be simply solved using a pattern. Sometimes, it only marks the starting point for an advanced diagnosis. Here, the language-aware debugger and sophisticated visualization features improve the error search over the usage of a standard debugger.

2.2.3 Debugging Support

Dynamic program slicing is a useful technique to reduce debugging effort. The designer can focus on the subsets of the system model only relevant for the observed error (see Section 1.1.2 on page 73). But how can we compute dynamic slices? Most algorithms base upon backward analysis using data and control dependencies. The program is instrumented and a trace is recorded for a concrete run. The trace contains all variables that were written and read for each statement. Moreover, for each control statement a control predicate variable is created. The control statement writes this predicate variable while for all controlled statements a read access is recorded. We want to compute the dynamic backward slice $S_d^B(v, q, x)$ for a variable v at execution position q on program input x. Therefore, the algorithm goes backward through the trace. It searches for the statement, v was last written and put it into the dynamic slice. If the value of v is calculated using some variables v_i, the algorithm recursively proceeds for each v_i until no more calculated variable references could be found.

Example 12. *Let us compute the dynamic backward slice for variable c in line 7 of the example program from Figure 4.2 using the corresponding (partial) trace:*

```
Trace                Read       Write
--------------------------------
1 p = 0;                        p
2 i = 1;                        i
3 c = 1;                        c
4 n = read();                   n
5 g = read();                   g
6 if (g = 1)         g          p6
7      c = 10;       p6         c
...
```

This statement is controlled by the predicate variable in line 6, and thus the statement is put into the slice. In line 6 variable g is read and last written in line 5. So, the dynamic slice is $S_d^B(c, 7, \{3, 0\}) = \{6, 5\}$.

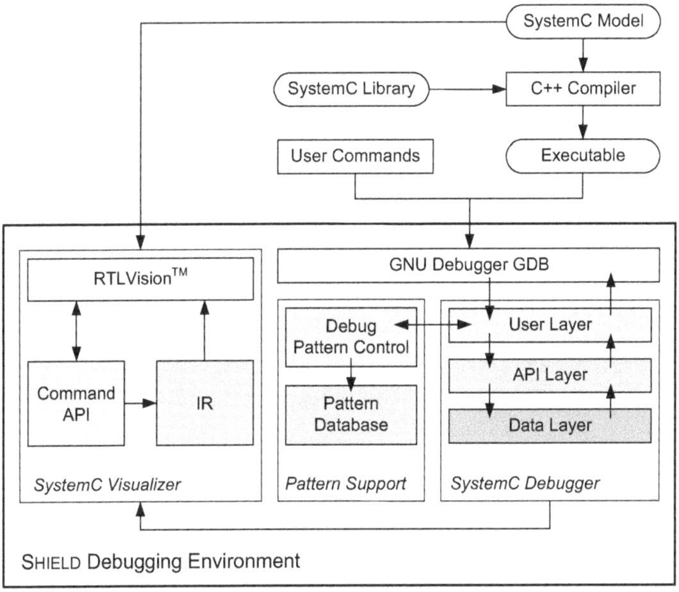

Figure 4.5: General architecture of SHIELD

3 HIGH-LEVEL SYSTEMC DEBUGGING

The integrated debugging environment SHIELD implements the debug flow as presented in Figure 4.4. There, the flow is adapted to the specific needs of debugging SystemC designs. First, this section presents the general architecture of SHIELD. Next, the introduced debug levels and the associated components are detailed. Finally, important implementation issues are discussed.

3.1 General Architecture

Figure 4.5 shows the general architecture of SHIELD. To enable debugging, the original SystemC description is compiled using a standard C++ compiler and the actual OSCI SystemC kernel library. Then, the created executable can be simulated inside the debugging environment. SHIELD consists of three major components.

The *SystemC Visualizer* cares for the visualization of SystemC designs. It statically analyzes the design and generates an intermediate representation. If a debug session is started, this representation is used to render the model inside the graphical front-end. The tool RTLVision™ from Concept Engineering is used for this purpose, exemplarily. Through the *Command API* the visualizer communicates with the debugger kernel where the exchange of data is done using a proprietary protocol based on socket communication.

The *SystemC Debugger* bases upon the Open Source debugger GDB and extends the functionality of GDB in a non-intrusive fashion (see Section 2.1 on page 77). That means, the solution works with any GDB version ≥ 6.3, the OSCI SystemC kernel reference implementation version $\geq 2.0.1$ [OSCI], and avoids any changes of debugged design code. To start debugging, the created executable is read into SHIELD and the simulation is started. After passing the SystemC elaboration phase successfully the debugger waits for user commands to debug the current design. Moreover, the debugger controls and alters the design visualization. The SystemC debugger has a layered architecture to ease the integration of new debugging functionality into GDB without the need for patching its sources.

The *Pattern Support* component is closely linked with the actual debugger. Therefore, the *Debug Pattern Control* unit monitors a concrete simulation run. If a specific failure symptom has been occurred during simulation, potential applicable debug patterns will be proposed. Otherwise, the designer manually chooses a pattern that is taken form the *Pattern Database*.

3.2 Debug Pattern Support

The strategy level (see Figure 4.3) is supported by the *Pattern Support* component (see Figure 4.5). Goals of the implementation are a simplified usage and a widely automated debugging process in order to relieve the user from standard debugging tasks. First, this section describes two important features. Next, a catalog of debug patterns is introduced.

3.2.1 Scenario-Based Guidance

Each debug pattern is represented by specific sequences of debugging commands forming a so called scenario. A call of the pattern name initiates a new pattern execution. The user is guided through the scenario by calling the command *nps* (*next-pattern-step*) which proposes a particular debug action to be done at next depending on the current situation. The *eps* (*end-of-pattern*) command finishes pattern guidance. Simultaneously to the text-based control, a flow chart documents the advance while executing the debug pattern. The scenario-based guidance significantly enhances the usability of patterns and enables the efficient application of the provided system-level debugging features with minimal learning effort.

3.2.2 Partially Automated Process

Sometimes, the currently occurred failure symptom allows SHIELD to automatically propose matching debug patterns. For this purpose, different information sources can be used:

- a reached simulation or debugger state,
- a formerly performed user action, or
- a processed history log of released events or triggered processes.

Aborting the current simulation run by pressing Ctrl-C since the debugger seems to hang in one of its processes is a typical situation asking for debugging support. Here, the environment suggests to use one of the LOCK patterns to find the defect causing the process to hang (see Section 3.2.3 on page 86). Another failure symptom is indicated by an event which triggers multiple processes in the same delta cycle. Then, the call of a debugging command lets proceed pattern guidance automatically to the next step.

Example 13. *The following GDB terminal log illustrates the scenario-based guidance in case of the TIMELOCK pattern used on an example SystemC design. Due to an obviously hanging simulation, the user interrupts it. Now, SHIELD tries to automatically diagnose the lock situation. Since the debugger does not report any process ID, the user manually checks for the TIMELOCK pattern by calling dp_timelock. So, a process is returned that is pending potentially. Simultaneously, the pattern guidance proposes automatically a usage of the lst command to examine the source code of the pending process. The next pattern step suggests to check for faults where a step-wise simulation could ease debugging:*

```
Program received signal SIGINT, Interrupt.
(gdb) dp_timelock
*** following process seems to hang
top.i_device._rx_tx (thread ID: 4)
*** TIMELOCK debug pattern activated
*** call lst 4 to examine source code of pending process
(gdb) nps
*** Check code for faults and proceed simulation step-wise.
(gdb)
```

3.2.3 Supplied Debug Patterns

Currently, the debug pattern catalog consists of seven patterns. These patterns were developed in close co-operation with SystemC designers. Each pattern represents a typical debug situation often occurred in daily work:

- COMPETITION. This pattern guides the user in the detection of competitive situations caused by nondeterministic process execution potentially resulting in erroneous behavior.
- TIMELOCK. The TIMELOCK pattern helps the user to find defects resulting from infinitely looping processes that cause no advance in simulation time (simulation freeze).
- DEADLOCK. When two or more processes are waiting for another, e.g. to release a shared resource, this pattern proposes a procedure to find the deadlock problem.
- LIVELOCK. This pattern provides a scenario to handle problems where two or more processes are working together, constantly changing their states but never coming to an end.
- OVERFLOW. A thread stack overflow causes a simulation crash. This pattern helps to identify the threads where such an overflow has occurred.
- LOSTEVENT. The LOSTEVENT pattern provides a debug procedure to detect events that were missed because of multiple overwriting *notify* calls.
- PERFORMANCE. This pattern suggests a procedure to determine the bottleneck probably caused by often released events or multiple activated processes in design blocks.

Each pattern guides the user step-by-step to identify a defect that could have caused the observed failure symptom. Appendix B details the COMPETITION and the TIMELOCK patterns as two representatives of the pattern catalog.

3.3 SystemC Debugger

Debugging a SystemC design is characterized by several recurring actions. Each action describes a sequence of particular steps in the GDB debugger to acquire needed high-level information at system level. Based upon such actions, SystemC-specific system-level debugging commands have been defined and implemented as a non-intrusive extension of GDB. According to the requirements presented on page 77, two types of commands are provided,

Table 4.1: Selection of system-level debugging commands

Examining commands	
Static simulation information	
lsm	list all modules in the given hierarchy
lse	output all events instantiated in the design
lsb	list all bindings of the specified channel
Dynamic simulation information	
lpt_rx	list all trigger events of the given process w.r.t. a specific time stamp
lst	output the code line a given process is currently pending
lsp	output all [c]thread and method processes
Controlling commands	
ebreak	break on next invocation of any process that is sensitive to the specified event
pstep	break on next invocation of the given process
tstep	break on processes which will be active at the next simulation time stamp

i.e. *examining* and *controlling commands*. Currently, SHIELD provides 31 commands that support an efficient and easy error search at system level. Table 4.1 gives a selection of typical representatives.

As already shown in Figure 4.5, the SystemC debugger has a layered architecture. Due to the demand for a non-intrusive extension of GDB, all new system-level debugging commands are implemented on top of it. Recurring actions are encapsulated into user-defined commands composing the so called macro instruction set at the *user layer*. A macro instruction implements a desired functionality by using built-in GDB commands, e.g. examining the symbol table or the backtrace, and a set of auxiliary functions provided by the *API layer*. Auxiliary functions are C++ or script helpers that evaluate and process information supplied by the debug data pool representing the *data layer*. The data pool obtains data from three sources:

- redirected output of GDB commands, i.e. temporarily created log files,
- direct access to internal data structures of SystemC kernel classes, or
- a database holding preprocessed system-level debugging information collected during set up of a debug session.

3.3.1 User Layer

The user layer provides the interface to all SystemC-specific debugging features in terms of macro instructions. It comprises the macro instruction set that is implemented by particular GDB script files. Moreover, this layer contains GDB helper scripts to setup and initialize SHIELD.

Example 14. *The lsb command (see Table 4.1) presents the common implementation frame at the user layer as used for all new system-level debugging commands.*

```
define lsb
   if ($hd_elaborated)
      echo ---lsb: list all bound ports---\n
      call hd::list_bound_ports($arg0)
   else
      echo not elaborated yet\n
   end
end
```

3.3.2 API Layer

The API layer supports the implementation of new debugging commands at the user layer. It is divided into an *auxiliary function API* and a *database API.* The auxiliary function API comprises in addition to awk/shell scripts, particularly C++ functions which realize more sophisticated helper functionality. Scripts are normally used to straightforward process temporary log files. The database API supplies interface functionality to store data into and to retrieve data from the debug database.

Example 15. *Figure 4.6 illustrates the actions to implement the lsb command by using a UML sequence diagram. A call of this command invokes the function hd::list_bound_ports(const char*) provided by the auxiliary function API. This function retrieves the corresponding sc_interface instance (a channel) using the SystemC method sc_core::sc_find_object (using the OSCI SystemC 2.2.0 reference kernel implementation). If the specified channel could be found, hd::list_binding(sc_interface*) is called. This database API function fetches the static binding information from the debug database and formats them accordingly for output.*

3.3.3 Data Layer

Three sources compose the data pool at the data layer supplying either static or dynamic simulation information.

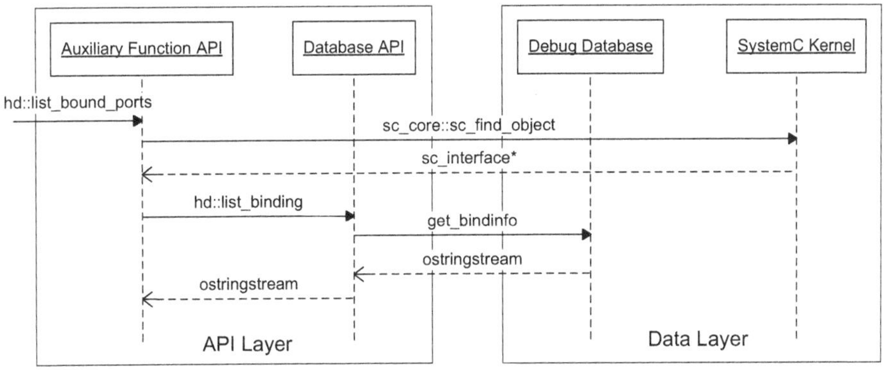

Figure 4.6: *lsb* command at the API layer

Temporary log files will be created by redirecting the output of GDB commands, e.g. to get the ID of the current thread the simulation has stopped, or the actual backtrace. These log files provide dynamic simulation information only accessible at debugger side.

The *SystemC kernel* provides some basic introspection capabilities useful for retrieving design and runtime information. Various global classes allow to query static simulation information, such as port, module, channel, or SystemC object registries.

Example 16. *The object hierarchy of a SystemC design can be browsed using the following loop.*

```
sc_simcontext* c = sc_get_curr_simcontext();
sc_object* o = c->first_object();
while (o) {
   if(!strcmp(o->kind(),"sc_module")) {
      // module specific actions
   }
   else if(!strcmp(o->kind(),"sc_signal")) {
      // signal specific actions
   }
   ...
   o = c->next_object();
}
```

The simulation control, implemented by the kernel class `sc_simcontext`, encapsulates the simulation state. This class offers many valuable information such as runnable processes at the next delta cycle, or the delta event queue.

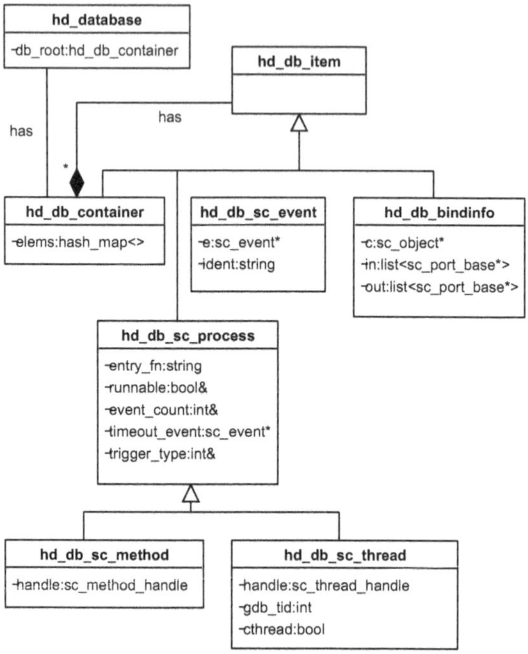

Figure 4.7: Class hierarchy of the debug database

During setup of a new debug session, static simulation information are logged and stored into the *debug database* using GDB. Here, particularly the ability of a debugger to fetch private class data is utilized. Otherwise, private data cannot be accessed due to missing public methods. Supplied debug functionality is based on four main information classes: event, binding, method and thread process information. Each class is represented by its own data type holding preprocessed, kernel-private, or special debug session data. Figure 4.7 depicts the UML class diagram of the debug database sketching the four information classes in the database class hierarchy.

Example 17. *The following lsb call retrieves the binding information for the channel 'i0_count_hier.count_sig' applied on an example application. There, the flow as sketched in Figure 4.6 is performed.*

```
(gdb) lsb "i0_count_hier.count_sig"
---lsb: list all bound ports---
bindings of channel i0_count_hier.count_sig
Driver:
   i0_count_hier.i_counter.outp    <sc_out>
Drivee:
   i0_count_hier.i_signal2fifo.inp   <sc_in>
```

Table 4.2: Selection of visualization debugging commands

Command	*Description*
vlsb	visualize the specified channel or event and all bounded components
vlsio_rx	highlight I/O ports of the specified module matching the given regular expression
vtrace	trace the given channel or port, record its value at each simulation time step until the specified time is reached and annotate the traced values as component label
vtrace_at	trace the given channel or port and record its value at the specified time steps, the recorded values are annotated as component label
vslice	visualize for the given slicing criterion each statement in the source code view that is part of the dynamic slice

3.4 SystemC Visualizer

To facilitate debugging of (complex) designs, SHIELD provides visualization capabilities based on the tool RTLVision™ from Concept Engineering. The visualization engine generates different views of the debugged design, supporting crossprobing and annotations of the visualized context. During a debug session the user has various possibilities to explore static and dynamic design information. A number of system-level debugging commands influence the graphical view (see Table 4.2). These commands are directly propagated to the SystemC visualizer. Being aware of the model structure, the visualizer assembles commands and maps SystemC components to the appropriate graphical symbols. Thus, RTLVision™ can be instructed to switch to specific parts of the design and to update signal values during execution. The following visualization features are available:

- annotating SystemC object names and values to signals and ports,
- hierarchical visualization of the module hierarchy,
- crossprobing between the graphical view and the source code,
- path fragment navigation,
- visualization of dynamic slices in the source code view, and
- module exploration with a highlighting of signals, modules, or ports.

The visualizer provides different views that allow to explore a SystemC design at arbitrary levels of detail. The *schematic view* shows modules as functional blocks that can be collapsed and expanded. Interconnecting wires represent some kind of interrelation between modules such as signals exchanging values between modules, synchronization dependencies between

```
131//
132tlm::tlm_generic_payload* ls_ctrl::get_next_data_word()
133{
134 Affects ls_ctrl.cpp:165 payload* pl = 0;
135
136  switch (state)
137  {
138    case WRITE:
139      // handle write bursts
140      if (!wr_bursts.num_available())
141        write_done.notify();
142
143      pl = wr_bursts.read();
144
145      if (rd_bursts.num_available())
146        state = READ;
147      else
148        read_done.notify();
149      break;
150
151    case READ:
152      // handle read bursts
153      if (!rd_bursts.num_available())
154        read_done.notify();
155
156      pl = rd_bursts.read();
157
158      if (wr_bursts.num_available())
159        state = WRITE;
160      else
161        write_done.notify();
162      break;
163  }
164
165  return pl;
166}
167
```

Figure 4.8: Dynamic slice for variable *pl* (code of the SIMD example)

different processes using events (notify/wait pairs), or transaction channels connecting sockets of TLM modules. The *cone view* limits the set of currently displayed objects to a specific path. Cone and schematic view are bidirectionally connected to the particular code lines in the *source code view*. The design hierarchy is represented as a compact tree-like structure which allows their fast browsing. Static and dynamic debugging information are shown by different colorings, info boxes, and labels, or output in the *debugger console*.

Dynamic slicing helps the designer to reduce debugging effort. Therefore, SHIELD instruments the code of the system model using a tool based on the REGATTA analysis framework (see Section 2 on page 40). This tool utilizes the dataflow analysis capabilities of REGATTA to extract read and write accesses of all variable occurrences. In case of an erroneous simulation run, a trace is created by the instrumented system model. The designer navigates in the source code view to the variable occurrence of interest. Then, based on the recorded trace the dynamic slice is computed. The slice is visualized by highlighting all statements that belong to that slice.

Example 18. *Figure 4.8 shows the dynamic slice for the variable pl in line 165 of the load/store controller implementation of the SIMD design example* (see Section 3.4 on page 28). *All statements that belongs to that slice are marked.*

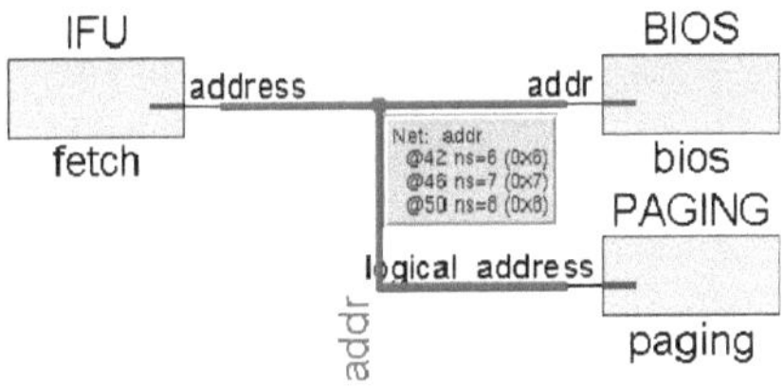

Figure 4.9: Visualization debugging command *vtrace_at*

Subsequently, the RISC-CPU design is used for further demonstration purposes of visualization capabilities. The CPU design is part of the OSCI SystemC v2.0.1 library package [OSCI].

Example 19. *Monitoring dedicated values during simulation is very helpful when the user does not exactly know what is going wrong and when the defect infection has occurred. In this situation, the vtrace_at command (see Table 4.2) helps the designer to trace the channel or port of particular interest. Figure 4.9 illustrates the tracing of the top-level signal 'addr' in the RISC-CPU design at three time stamps to check whether the right addresses are forwarded to the RAM. The concluding vlsb command displays the signal and its connection together with the annotated traced values:*

```
(gdb) vtrace_at "addr" 42000
(gdb) vtrace_at "addr" 46000
(gdb) vtrace_at "addr" 50000
(gdb) c
...
(gdb) vlsb "addr"
```

Example 20. *In case of a failure related to a specific channel, the user wants to get a quick overview about all channel connections. Here, the vlsb command (see Table 4.2) helps the designer to focus error search on the relevant modules. Figure 4.10 sketches the visualization output after calling vlsb with two top-level signals of the RISC-CPU design in order to check the right port binding:*

```
(gdb) vlsb "ram_cs"
(gdb) vlsb "next_pc"
```

3.5 Implementation Issues

An important requirement for twhe implementation of SHIELD was the demand for a non-intrusive solution. Hence, patches of the SystemC kernel, the GDB debugger, and the code should be avoided (see Section 2.1 on page 77).

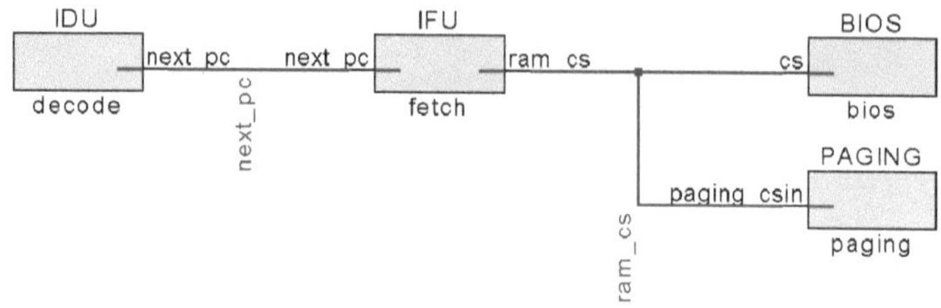

Figure 4.10: Visualization debugging command *vlsb*

Figure 4.11: Preloading a SystemC kernel method

A first implementation approach collected the required system-level debugging information using a *pure non-intrusive* procedure. Here, hidden breakpoints in the SystemC kernel register and trigger particular data collection actions, e.g. to collect created SystemC processes. Experiments with complex real-world designs have shown that especially the setup phase of the extended GDB took an unacceptable long time. Here, the frequent calls of data assembly breakpoints have downgraded the simulation performance.

So, an improved, more *relaxed non-intrusive*, approach was developed. The idea is to reduce the number of data assembly breakpoints while moving their functionality into the kernel methods where the breakpoints were formerly set. Normally, one has to patch the particular methods to create callbacks forwarding required debugging information to SHIELD. To remain kernel patch-free, library interposition was used instead. A preloaded shared library contains the overwritten SystemC kernel methods. Additionally, the original implementation is extended by a callback into the debugging environment. A specific environment setting instructs the dynamic linker to use the preloaded library before any other when it searches for dynamically loaded functions. Figure 4.11 illustrates this principle in case of the method `create_thread_process` of the class `sc_simcontext`. A callback forwards all information about a created thread to SHIELD. Since preloading works only for non-inlined class methods, minor transparent changes to the SystemC kernel source code were necessary. So, some inlined methods were moved from header to implementation files such as the constructor of the `sc_event` class. Using the new relaxed non-intrusive approach has sped the debug session setup times up to 62.4 times compared to the pure non-intrusive variant (see Table 4.3).

4 EXPERIMENTAL RESULTS

This section summarizes different results obtained with SHIELD while debugging flawed simulation runs of different SystemC designs. The first experiment shows an exemplary debug session of the SIMD data transfer example (see Section 3.4 on page 28). It demonstrates how the proposed debug flow together with the supplied debugging features help the designer in a systematic error search. The second part considers some experimental results that were obtained while debugging different industrial designs.

4.1 SIMD Data Transfer Example Continued

Initially, the modules of the SIMD design have been checked by static analysis (see Chapter 3). Now, the example is compiled and simulated the first time. To verify the correctness of data transfer operations, simple logging statements are used. The evaluation of the logs has shown that read and write data transfers wrongly start at the same time. So, the data transfer operations compete for the access to the data bus. Obviously, a failure is found and a SHIELD debug session is started to find the failure-causing defect. The observed failure symptom matches the symptom which is described by the COMPETITION debug pattern (see Appendix B). According to Figure 4.4, this pattern is selected and activated to start debugging at the strategy level:

```
(gdb) dp_competition
*** COMPETITION debug pattern activated
*** detect competitive situation
```

To get a better insight into the current design structure, the SystemC visualizer component of SHIELD is applied. Using the *vlsb* command (see Table 4.2), the internal structure of the processor unit is displayed in the *cone view*. This comprises the visualization of three thread processes and their synchronization dependencies using SystemC events. The thread process `core_i.punit.dispatch` is the central control unit. It synchronizes the data transfer processes `core_i.punit.read_data_block` and `core_i.punit.write_data_block` using the two SystemC events `e_read_grant` and `e_write_grant`. These processes initiate a new read or write transaction and report the completion by using the acknowledge events `e_read_ack`

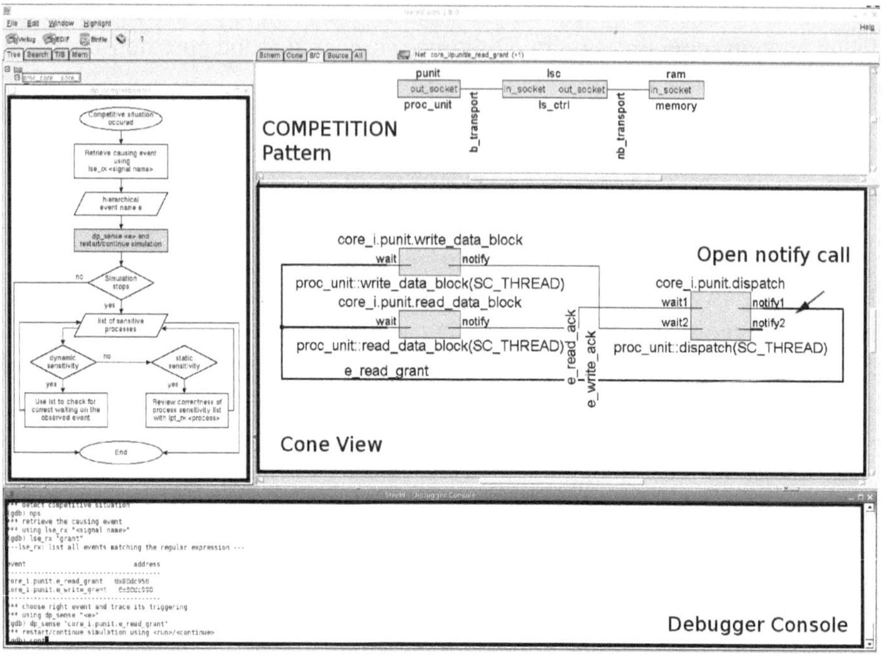

Figure 4.12: SHIELD screenshot of a SIMD debugging session

and e_write_ack. As can be seen in Figure 4.12, the e_write_grant event does not notify any process. Instead, both data transfer processes seem to wait for the e_read_grant event. This observation could already cause the examined failure. To validate this assumption, the COMPETITION pattern is further processed:

```
(gdb) nps
*** retrieve the corresponding event
*** using lse_rx "<signal name>"
```

As proposed by the *nps* command, the *lse_rx* debugging command is used to retrieve the hierarchical name of the interesting read-grant event. After calling *lse_rx*, the pattern guidance automatically jumps to the next debugging step (in the flowchart). Then, this step proposes to apply the *dp_sense* command on the event of interest and continue simulation. This particular situation is shown in the screenshot of Figure 4.12. Now, the *dp_sense* command is checking continuously whether more than one process is triggered by the read-grant event, and if so, the simulation would be stopped. As result, *dp_sense* reports two sensitive processes becoming active simultaneously. This situation corresponds to the visualized output already shown in Figure 4.12:

```
*** dp_sense 'core_i.punit.e_read_grant'
*** Check competitive situation between sensitive processes
  in module core_i.punit
     core_i.punit.write_data_block  <dynamic>
     core_i.punit.read_data_block  <dynamic>

*** breakpoints in sensitive processes
  breakpoint at thread 4  -> core_i.punit.write_data_block
  breakpoint at thread 5  -> core_i.punit.read_data_block
```

Due to the detected dynamic sensitivity of both potentially competitive processes, the debug pattern proposes to use the *lst* command. This command checks the correct process sensitivity by investigating the code the process is currently pending. So, *lst 4* depicts the defect in line 91. Here, the process waits for a read grant signal in the write-data-block. This seems to be a (typical) *copy-paste* failure from the read-data-block method:

```
(gdb) lst 4
---lst: list source a [c]thread/method is currently in---
process core_i.punit.write_data_block is currently
  at proc_unit.cpp:91
  in proc_unit::write_data_block
89        while (true)
90        {
91          wait(e_read_grant);
92
93          // how much data should be written
--------------------------------------------------------
```

Summarized, the failure-causing defect could be detected by a systematic procedure at the strategy level. Using the COMPETITION debug pattern in connection with *three* system-level debugging commands is sufficient for defect analysis. The visualization of the relevant design parts further eases debugging. So, the visualized process dependencies have been early pointed out the faulty implementation.

4.2 Industrial Examples

This section considers experimental results which were obtained by exploring different SHIELD features on various industrial SystemC designs. Furthermore, the debugging features of SHIELD are contrasted with a commercial solution, namely the CoWare SystemC shell.

Table 4.3: SystemC debugger setup times

Test	*Setup time*		
	C++ debugger	*Pure non-intrusive*	*Relaxed non-intrusive*
Design A[a]	<1 s	2.25 min	1.2 s
Design B[b]	~10 s	26 min	25 s
Design C[c]	~10 s	>40 min	31 s

a. *Test system:* AMD Opteron™ 248 processor @2200 MHz, 3 GB RAM
Test design: multiple instances of a simple producer/consumer application
Design characteristics: #thread processes – 204, #method processes – 1010, #sc_events objects – 3638

b. *Test system:* AMD Athlon™ XP processor @1800 MHz, 3 GB RAM
Test design: bus interface controller for various protocol implementations
Design characteristics: #thread processes – 29, #method processes – 56, #sc_events – 306

c. *Test system:* see b
Test design: design of a serial interface
Design characteristics: #thread processes – 31, #method processes – 136, #sc_events – 701

4.2.1 Efficiency Discussion

For a successful deployment in an industrial design flow, the SHIELD environment and especially the SystemC debugger should offer a competitive performance and productivity if compared to a standard C++ debugger.

Table 4.3 presents the setup times, the debugger needs to be ready comparing the non-intrusive implementation variants (see Section 3.5 on page 93). The relaxed non-intrusive approach combines flexibility and the demand for unpatched sources with a setup time that competes well with the standard GDB implementation. Moreover, the table shows the bad performance of the pure non-intrusive approach that prevents any productive work.

Table 4.4 illustrates the performance of the *vtrace* debugging command (see Table 4.2) while tracing a different number of signals on three simulation runs of the RISC CPU design [OSCI]. So, the observation of 750,000 time points over 125 signals leads to a slow down of factor 4 compared to a trace-free simulation. The tracing of 50 signals increases the simulation time only about 80%.

Rating the efficiency of the SystemC debugging features is difficult. To get an idea, Table 4.5 compares the effort the user has to invest if system-level debugging commands are replicated by a sequence of standard GDB commands. Note that this is impossible for half of the functionality. For the other half, the user has to have at least a deep understanding and a detailed knowledge of the SystemC simulation kernel. More often, it requires additional functionality and data collected during a debug session.

Table 4.4: Exemplary performance slow down due to tracing[a]

#Traced signals in the RISC CPU design	Slow down over simulation time (#observed time points)		
	1,000 ns	*2,000 ns*	*3,000 ns*
0	1.0	1.0	1.0
5	1.3 (10,000)	1.3 (20,000)	1.4 (30,000)
50	1.8 (100,000)	1.8 (200,000)	1.8 (300,000)
75	2.3 (150,000)	2.6 (300,000)	3.0 (450,000)
100	3.0 (200,000)	3.2 (400,000)	3.6 (600,000)
125	3.2 (250,000)	3.9 (500,000)	4.0 (750,000)

a. *Test system:* Intel Centrino Duo T2400 @1830 MHz, 1 GB RAM
Test design: Design is part of the OSCI SystemC v2.0.1 kernel package
Design characteristics: #thread processes – 9, #method processes - 3, #sc_events – 173

Table 4.5: Debugging effort in SHIELD and GDB

#HLD commands	*#GDB commands per system-level command*	*Remarks on equivalent GDB functionality*
6	1 to 9	fully automated with canned command sequences, i.e. a number of GDB commands
10	dynamically/at least 4	partially automated with manual user intervention, needed to handle dynamic information
15	n/a	additional functionality required, i.e. helpers to retrieve, evaluate, and buffer system-level information

4.2.2 SHIELD in Practice

In the following, the SHIELD debug flow is compared with a conventional C++ based debug procedure. Here, a bug should be detected in a SystemC test bench running a co-simulation between Verilog and SystemC.

4.2.2.1 Debug Problem

During running a test case, an interface bus of the actual design under test (DUT) had a value contention. It seems that there are two processes concurrently driving data onto the bus.

The first process in the test bench is known. It drives an initialization value after reset using the thread process *tb.bif.fsm._reset()*. This thread is sensitive to the negative edge of the reset signal *fsm_rst_l*, being a low active input to the DUT. So, the second yet unknown process shall be detected. Since the signal event on which the problem occurs is known, a tracing of the corresponding SystemC event *tb.bif.fsm_rst_l.m_negedge_event* could help to isolate the failure cause.

4.2.2.2 SHIELD Debug Procedure

The assumption of two processes that concurrently drive data onto the interface bus, motivates the application of the COMPETITION debug pattern (see Appendix B). So, the erroneous test case is rerun in SHIELD. Since the problem-causing signal is already known, the *dp_sense* command is called with the particular signals negative edge event:

```
(gdb) dp_sense "tb.bif.fsm_rst_l.m_negedge_event"
*** COMPETITION debug pattern activated
*** restart/continue simulation using run/continue
(gdb) cont
```

On stop of *dp_sense*, it gives us two thread processes sensitive to it:

```
*** dp_sense 'tb.bif.fsm_rst_l.m_negedge_event'
*** Check competitive situation between sensitive processes
  in module tb.bif
     tb.bif.if_bfm._update <static>
     tb.bif.fsm._reset <static>

*** breakpoints in sensitive processes
  breakpoint at thread 21 -> tb.bif.fsm._reset
  breakpoint at thread 16 -> tb.bif.if_bfm._update
```

According to the debugger output, the second sensitive process is *tb.bif.if_bfm._update*. Due to the static sensitivity, the COMPETITION pattern proposes to review the process sensitivity list using the *lpt_rx* debugging command (see Table 4.1), at next:

```
(gdb) lpt_rx 16
thread process sensitivity list
-------------------------------
tb.bif.if_bfm._update
  <dynamic> tb.bif.ch_m._update.m_value_changed
  <static> tb.bif.fsm_rst_l.m_negedge_event
  <static> tb.bif.fsm_tx_w.m_posedge_event
```

As can be seen, the sensitivity list falsely includes the reset signal, which is not desired and turns out to be an environment bug.

While debugging at the strategy level, only *three* system-level debugging commands (including the *next-pattern-step* command) are necessary to figure out the failure cause. In contrast, a conventional debug procedure is much more expensive, as sketched in the following.

4.2.2.3 Conventional Debug Procedure

A conventional debug procedure could only use standard GDB debugger commands. To find the second yet unknown process colliding with the reset process, a breakpoint would be set into the unique driver function. This breakpoint is invoked by every module that wants to stimulate the interface bus.

At any stop at this breakpoint, the designer had to trace back the invoking module, e.g. with the *up* command in GDB. This turns out to be a time consuming task, potentially ending in different modules not involved in this specific issue. The procedure can be simplified if the knowledge about the problem-causing SystemC signal event is used. Since system-level debugging commands are not available, a conventional C++ breakpoint has to be set at the proper SystemC kernel method that is called if an event is triggered. To stop only at the searched event, a breakpoint condition has to wait for the particular event object. Subsequently, all processes sensitive to this event are retrieved and checked whether they are correctly triggered. Again, this procedure is a very time consuming task which needs a lot of user input and also an intimate knowledge of the SystemC kernel implementation. Usually, this knowledge cannot be expected to be available by a normal designer.

4.2.3 SHIELD Debugger vs. CoWare SystemC Shell

In the following, the current implementation of SHIELD's SystemC debugger (see Section 3.3 on page 86) is compared with the commercial CoWare Platform Architect SystemC development environment (Product Version V2007.1.1) [CoWare], precisely the SystemC shell *scsh*. Just like the SHIELD environment, the *scsh* tool provides a debugger interface at command line.

Table 4.6 contrasts the most important debugging features in SHIELD with the corresponding *scsh* features. As can be seen in the table, both tools provide comparable debugging features and base upon GDB. One of the main differences between the environments is the integration of GDB. While SHIELD extends an arbitrary GDB release in a non-intrusive, patch-free manner, CoWare's *scsh* fully integrates a specific GDB version into a Tcl/Tk scripting engine. Both tools offer breakpoints at system level. However, SHIELD breakpoints are set on all activated processes that are triggered by a certain event, e.g. a signal value change, or entering the next delta cycle.

Table 4.6: Comparing SHIELD debugger and CoWare's *scsh*

Feature description	*SHIELD*	*scsh*
GDB debugger based	yes	yes
GDB extension	use GDB extension concept	Tcl/Tk based
SystemC project support	no (only GDB built-ins)	yes (build and run simulations)
system-level breakpoints		
at the initialization phase	no	yes
at simulation time advance	breakpoint at each activated process	only stop
at delta cycle advance	breakpoint at each activated process	no
on SystemC event/signal/port	breakpoint at each activated process if event has occurred	breakpoint at a specific location if event has occurred
on process	yes	yes
print/display signals/variables	signal tracing feature/built-in GDB variable value printing	specific commands
navigate design hierarchy	similar features to list different design information - document only differences, subsequently	
list whole object hierarchy	no	yes
list transfers/attributes of	no	yes
display item information	no	yes
list process trigger	yes	no
list all events in the simulation	yes	no
tracing features	signal, port, and process trigger tracing	sophisticated tracing of arbitrary traceable objects
visualization interface	specific commands to alter the visualization front-end	no (only debug shell, visualization integrated into IDE)

In contrast, in *scsh* the designer sets a breakpoint at a specified location to stop simulation whenever the given event has occurred, e.g. an elapsed number of clock cycles. The different procedures have advantages and disadvantages.

So, SHIELD enables a more communication-centered debugging while *scsh* allows to stop simulation at more specific situations. Both tools provide similar features for a navigation inside the design hierarchy. Here, *scsh* supports a more detailed listing of particular SystemC objects while SHIELD assists debugging through additional dynamic simulation information. The tracing features are better supported in CoWare's *scsh* while SHIELD provides a port to alter the visualization front-end from the console.

5 SUMMARY AND FUTURE WORK

A debug flow that supports systematic debugging of ESL designs has been introduced in this chapter. A three level strategy allows an efficient observation of simulated system models. It guides the designer in a fast, easy, and simplified debug procedure. The SHIELD debugging environment implements the proposed debug flow for SystemC. It assists designers in simulating, debugging, and visualizing their SystemC models. Thus, the designer gets a quick and concise insight into the static structure and the dynamic behavior of the model without the burden of gaining detailed knowledge of the underlying SystemC simulation kernel. A special feature of SHIELD over all other available approaches and tools is its non-intrusive implementation that avoids any patches of the used debugger. Only minor transparent changes to the OSCI SystemC reference kernel implementation are required, especially to enhance debugging performance. Hence, SHIELD can be easily integrated into an existing industrial design flow. Providing concrete debug strategies in terms of debug patterns help the user to focus on a higher level of abstraction joined with a minimal learning effort. So, common sources of typical SystemC errors can be systematically debugged using a particular debug pattern. Especially the novice user can exclude and fix many defects without the need to consult an expert. The experienced user who does not need pattern guidance anymore has the benefit of a comprehensive support. Here, the partially automated detection of erroneous situations can accelerate the debug process.

Practical experiments have demonstrated the efficiency of SHIELD while debugging various industrial SystemC designs. SHIELD has shown a setup time which is only three times slower at the maximum, and thus competes with the original GDB implementation. GDB is a de-facto standard debugger for C and C++ that has proven its feasibility to handle very complex software programs, e.g. as default part of the Eclipse CDT plug-in [CDT]. Moreover, SHIELD supplies system-level debugging commands whereas 50% of these commands do not have a direct correlation to standard GDB features. The other half of commands needs four GDB commands on the average to replicate the corresponding debugging functionality at system level. So, a defect in the SIMD data transfer example could be found in a few steps using a debug strategy combined with a dedicated visualization of the corresponding module. Consequently, it could be assumed that SHIELD enables a more efficient debugging of SystemC descriptions than using a standard C++ debugger.

Future work will improve the provided debugging and exploration functionality especially regarding an explicit TLM support. One of the main goals is to fit the debugging environment to the specific needs of the application being developed, e.g. CPU design.

Chapter 5

Learning about the Design

A single simulation run can be a valuable source of information to detect many errors in the system model. However, summarizing multiple runs to get a general abstraction that holds for all concrete runs offers completely new debugging opportunities. So, new and different aspects about the design can be extracted. Such an *induction technique* is introduced in this chapter (see Figure 5.1). In general, two main approaches can be distinguished.

The first approach uses multiple program runs to detect *abnormal* behavior, i.e. behavior that deviates from the average whereas these abnormalities, also anomalies, help to locate errors. Although, abnormal behavior does not necessarily indicate a defect as in case of incorrect behavior, it is usually a very good indicator to start your debug session. Anomalies are represented by certain properties that can be inferred from erroneous program runs.

The second approach uses multiple program runs over passing tests to infer likely specifications that hold for all runs. These inferred properties (also invariants) can be converted for instance into assertions that check for abnormal program behavior during subsequent program runs.

In the first part of this chapter a general overview about induction techniques is given. These techniques are used to determine a general abstraction of the analyzed program or the design description. Subsequently, a new methodology is described that generates complex properties over multiple simulation runs of a system model.

The second part of the chapter introduces DIANOSIS (Dynamic Invariant ANalysis On SImulation TraceS) – a methodology and tool that infers complex properties from a given simulation trace using two analysis phases. In the first phase, a set of predefined properties is hypothesized over the trace. During the subsequent second phase the previously found properties are combined to new, more complex candidates. These candidates are checked on the simulation trace once again. "Surviving" property candidates are recombined until no more valid properties can be created. Then, the extracted functional behavior of the design is presented to the verification engineer for manual inspection and subsequent use.

F. Rogin and R. Drechsler, *Debugging at the Electronic System Level*,
DOI 10.1007/978-90-481-9255-7_5, © Springer Science+Business Media B.V. 2010

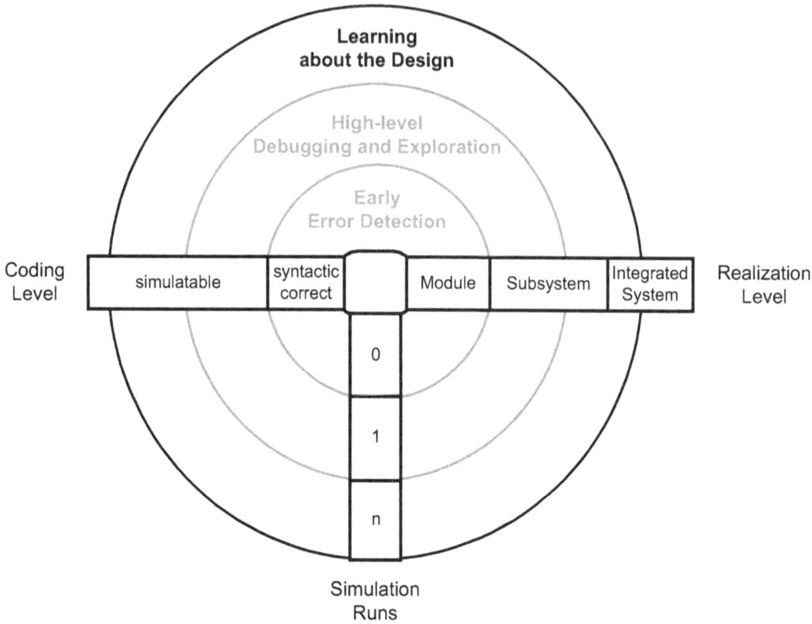

Figure 5.1: Learning about the design using an induction technique

1 INDUCTION TECHNIQUES IN A NUTSHELL

In this section induction techniques and their role to learn different aspects about a system are discussed, at first. Next, an overview about related work is given.

1.1 Overview

The increasing computing power has made dynamic analysis a promising approach in system design and software program verification. Different automated induction techniques infer abstractions in terms of properties using multiple program runs. Such properties could help the designer to learn important design characteristics or to detect erroneous behavior. Hence, two fields of application can be distinguished:

- *Inferring a likely characteristic.* Here, a (large) set of passing program runs is used to abstract behavior that is common for nearly all runs. Examples are a specific call order of API functions of a software package or the correct locking/unlocking of shared resources in multi-threaded applications. So, this approach tries to infer commonalities between multiple program runs that generally hold. The more runs are analyzed the more precise are the obtained results. The found common behavior can be converted into run-time checks (also assertions) to detect anomalies during later program runs. A flagged assertion indicates a situation where the program characteristic deviates from the "usual" behavior, seen so far. There, the user has to interactively decide whether this anomaly is a new, not previously observed correct fact or a failure.

- *Detecting abnormal behavior.* In contrast to inferred likely program characteristics, abstractions extracted from failing runs describe erroneous behavior. An example is a missing mutex unlock of a shared resource in a multithreaded application that causes a program deadlock. Comparing correct and wrong abstractions can help the user to locate defects that had caused the observed failure.

Both approaches need a suitable database storing data of multiple program runs for analysis. Moreover, the analyzed program must be extended to collect the required data. Data collection can be provided by automated instrumentation techniques. An instrumented program logs data that allow to keep track of executed program statements, or all values a variable holds during run time. To ease instrumentation, different frameworks and tools especially in the software domain are supplied such as Valgrind or Log4J for Java. In spite of the framework support, the collection of specific information often needs some programming expense. Instrumentation tools for C++ can be also used to instrument SystemC code. However, these tools cannot collect data specific for SystemC. SystemC provides a built-in concept that was taken over from hardware design. The designer can initiate a tracing of specific signal values to create a simulation trace in the VCD standard format to be visualized in a waveform viewer, later.

The larger the amount of data is, the higher the precision of the induced abstraction can be. A promising approach is the collection of data under real operating conditions, instead of generating a large set of synthesized random runs. Running a program under real-world conditions produces a high number of runs with a large variety of real-life scenarios. Nonetheless, collecting data in the field is problematic:

- A collection of arbitrary data could violate the privacy of the user.
- A collection of a large amount of data could negatively impact the performance of the application. This impact is critical for real-time applications.

In general, induction techniques require a suitable instrumentation approach and a large set of test cases creating a sufficient dataset. There, the ever increasing computing power and ongoing research progress improve the precision and potential of these techniques.

1.2 Related Work

The automatic generation of properties or specifications in the hardware as well as in the software domain is subject of numerous works.

The dynamic invariant detector Daikon [PE04] reports likely invariants found in programs for instance written in Java, C/C++, or Perl. Daikon searches a preset catalog of mainly algebraic invariants from program executions. It validates hypotheses against each execution trace of the instrumented program and thus relies on high-quality traces. Csallner and Smaragdakis [CS06] propose an algorithm that supports interface and method overriding in the context of Daikon. A method comparable to Daikon is presented in [HCNC05] for the hardware domain. Here, a small number of predefined properties (e.g. one-hot coded buses, different handshake patterns) is inferred over a simulation trace whereas the properties are especially adapted to the specific needs of microprocessor designs. Found properties can be introduced in subsequent verification tasks. In contrast to the solutions presented above, the approach proposed in this book is not restricted to a predefined set of properties. Instead, the approach reuses already generated properties to infer more complex ones. Hence, more abstract information about the analyzed design can be obtained.

The principle of "least astonishment" is utilized by the approach presented by Isaksen and Bertacco [IB06]. The method extracts common behavior by means of transaction activity at any user-defined component interface of a digital hardware design. The results are presented to the verification engineer by protocol or transaction diagrams. Any design anomaly is reported as a difference to the common behavior constituted in a learning phase. A disadvantage of this approach is the limitation to control signals to gain usable information instead of generating properties over all possible signals as our solution does. Due to this limitation, for example, no statements about data-oriented behavior can be made. Moreover, the presented experimental results show that the obtained transaction diagrams can quickly reach a complexity that hampers its comprehension and validation. In contrast, substantial effort was made in our approach to improve the readability of generated properties and to limit their number.

In the software verification domain Engler et al. [EC+01] propose a comparable approach to the approach of Isaksen and Bertacco [IB06]. Engler et al. compare common behavior with the current observed behavior using a set of predefined templates. Hence, the programmer can infer likely faults by inspecting behavior which deviates from the norm. Our generation methodology shares this observation where expected but unseen properties can point to an erroneous design or test bench implementation. The main difference to this work is that Engler et al. use static program analysis. So, this approach cannot be applied to programs where the source code is not available. Our approach only relies on simulation traces that are created by a simulation run.

Arts and Fredlund [AF02] presented a work to improve design understanding of Erlang programs by an automatic analysis of program traces. Thus, program models from execution traces are obtained that are used for visualization and model checking purposes. Due to the regular nature of Erlang programs, design patterns (e.g. finite-state machines, client–server communication, event handling) can be searched on the trace. As a result, abstract state graphs can be created from the traced component. Our work has some similarities with this approach, especially in the case of searching recurring patterns on a trace. While [AF02] is limited to Erlang programs, our methodology works on arbitrary simulation traces that are provided in a common industry standard data format.

The principle of "specification mining" is discussed in many different work. A mined specification is typically a finite-state machine describing valid scenarios by means of instruction sequences in a software program. In [ABL02] a machine learning technique is used to mine specifications for application programming interfaces and abstract data types. First, so called scenarios are extracted from a program execution trace. Then, these scenarios are fed to a probabilistic finite automaton learner. Erroneous traces are handled with the help of human experts who have to decide whether a violation is a real bug or not. More recent work extend specification mining to handle longer real-world execution traces [YE04], more realistic imperfect traces [YE+06], or to achieve more precise results [WN05]. All approaches are settled in the software domain typically analyzing temporal properties such as execution sequences of functions at the interface border of program components. In contrast, our approach is able to deal with arbitrary signals of hardware designs and searches for recurring patterns between them. There, not only temporal relations are extracted. Nevertheless, several mining techniques and features can help to improve our methodology.

Anomaly detection is performed by the tool DIDUCE [HL02]. It reports violations for relatively simple invariants (e.g. read/write specific values) detected in Java programs. In [DF04], [FD04] a methodology is introduced that extracts arbitrary properties on a simulation trace of a hardware design. Different time relations between a selected set of signals are considered, rated, and formulated as a property using pattern matching techniques combined with a heuristic. Besides the choice of correlated signals, a time window in which properties will be searched has to be defined by the user. However, this normally limited window length prevents an efficient search for properties, especially in case of communication protocols where large window sizes are common. A number of work detect design patterns in software programs to improve program understanding. Pattern detection works either by analyzing structural aspects of patterns, e.g. [NS+02], [SO06], [Vok06], or behavioral aspects, e.g. [SO06], [Wen03].

Compared to our approach, recovered design patterns mainly give statements about the code quality in terms of maintainability, documentation, or extensibility, and support reverse engineering tasks. The recognition of specific software defects, such as race conditions or deadlocks, are examined by various approaches, e.g. [PS03], [SB+97], [YRC05]. The characteristic of all work is their particular adaptation to the particular needs of the checked properties. Hence, their application in other contexts is very limited.

2 AUTOMATIC GENERATION OF PROPERTIES

In this section a methodology for the generation of properties based on simulation traces is proposed, at first. The methodology is suited to infer a likely specification and to detect abnormal design behavior. Then, the generation algorithm is illustrated in detail. Finally, the extension of a system design flow incorporating the introduced property generation methodology is discussed.

2.1 Generation Methodology

The proposed property generation methodology bases upon simulation traces. Hence, it shares similarities with the work presented in [GD04b], [HCNC05], [IB06]. The design description is manually instrumented by the designer to create traces in the industrial VCD standard format. There, the design description can be written in any language and at each abstraction level. The only precondition is a support of the VCD format as provided by system level descriptions written in SystemC, or RTL descriptions written in Verilog.

The key contribution of the presented methodology is the generation of complex properties by combining already found properties to new ones. In the first phase, a set of predefined properties is hypothesized over the design behavior which is described by a given simulation trace. During the subsequent second phase the previously found properties are combined to new, more complex candidates. These candidates are checked on the simulation trace once again. Surviving property candidates are recombined until no more valid properties can be created. Then, the extracted functional behavior is presented to the designer for manual inspection. The advantages of a methodology that automatically derives (complex) properties are as follows:

- *Improved design understanding.* Inferred properties gain a new insight into the abstract design behavior. So, properties that deviates from the average can point to a defect and are a good starting point for debugging.
- *Tool support for the formulation of properties.* Derived properties are a starting point for formal verification, e.g. by converting them into assertions.
- *Identification of gaps in the test suite.* If an inferred property is not a general abstraction that holds for all simulation runs, it unveils behavior not exercised by the test suite.
- *Enhanced efficiency of the overall verification process.* The application of such a methodology improves quality and efficiency of the overall verification process.

2.2 Generation Algorithm

First, this section introduces some basic notions to achieve self-containment. Then, the general property generation algorithm is described. Finally, termination and complexity of the algorithm is discussed in short.

2.2.1 Preliminaries

A simulation trace T has a vector S of n signals $S = (s_1, ..., s_n)$ consisting of inputs, internal signals, and outputs. The algorithm searches for properties over all given signals independent of their directions. So, adding information about the directions of signals is not necessary which saves configuration effort and improves process automation. Moreover, the search space for properties is enlarged to any feasible signal dependency. So, for example, a back coupling from an output to an input can be identified. The value of a signal s_i at time t is denoted by $s_i[t]$. Since, hardware signals can be low- or high-active, a heuristic h is defined that estimates the polarity of a signal s_i at time t. Similar to the work presented in [HCNC05], the heuristic assumes that the signal value found in a smaller number of clock cycles is the active polarity:

$$h(s_i[t]) = \begin{cases} 1 & \text{if } s_i[t] = \text{active at } t \\ 0 & \text{if } s_i[t] = \text{inactive at } t \end{cases} \tag{5.1}$$

Hence, the heuristic presumes that a digital circuit is a longer time inactive than active when considering its overall lifetime. This assumption implies that the used simulation trace also reflects the real-world behavior. Otherwise, if a signal is more often active than inactive its polarity would be wrongly derived. Consequently, wrong design behavior could be inferred. Using the heuristic h a signal state $s_i[t]$ at time t is encoded into a signal activity $a_{s_i}[t]$.

Definition 11. *A simulation trace T of length* t_{cyc} *is given by a tuple* $(S, (v[1], \ldots, v[t_{\mathrm{cyc}}]))$ *with*

- $S = (s_1, \ldots, s_n)$ *is a vector of n signals, and*
- $v[t] = (a_{s_1}[t], \ldots, a_{s_n}[t])$ *gives the activities of these signals at time t.*

2.2.2 Algorithm Description

Figure 5.2 presents the property generation algorithm. The algorithm starts from

- a simulation trace T,
- a regular expression string rx which restricts the number of examined signals in S to an interesting subset,
- a set of user-defined basic and complex checkers verifying the corresponding properties with $C = C_{basic} \cup C_{complex}$, and
- information about reset- and clock-signals, i.e. their names s_{rst}, s_{clk}, and the reset phase finish time t_{rst}.

As a result the algorithm delivers a set P containing all valid basic and complex properties found on T. Additionally, several parameters are gathered during the generation process which specify each found property. A property parameter is for instance the maximum clock cycle between a request and an acknowledge signal in case of a signal handshake. A *valid property* confirms the checked "characteristics of the participating signal(s)" such as the handshake pattern between two signals. Using this information and the collected parameters, an expression in a particular property language is given to the user, e.g. in terms of an SVA, or a PSL expression.

The algorithm is composed of two phases. During the first phase (lines 1 to 9) predefined basic properties are inferred over the simulation trace. In a first step, a set of property candidates P_{cand} (line 2) for all signals and their combinations over all requested basic checker types C_{basic} is created. Then, the validity of basic property candidates is checked at each time

```
Require: T = (S, (v[1], ... ,v[t_cyc])), [rx], C, s_rst, t_rst, s_clk
Ensure: P with p ∈ P is valid on T
1   {phase 1 - infer basic properties}
2   P_cand = create_basic_cand(S, C_basic, s_rst, s_clk, rx)
3   for all time steps t_rst ≤ t ≤ t_cyc do
4       for all p ∈ P_cand do
5           if check_p(v[t]) = false then
6               P_cand = P_cand \ p
7           end if
8       end for
9   end for
10  P = P_cand
11  {phase 2 - infer complex properties}
12  repeat
13      P_cand = create_complex_cand(P, C_complex)
14      for all time steps t_rst ≤ t ≤ t_cyc do
15          for all p ∈ P_cand do
16              if check_p(v[t]) = false then
17                  P_cand = P_cand \ p
18              end if
19          end for
20      end for
21      P = P ∪ P_cand
22  until P_cand = ∅
```

Figure 5.2: General property generation algorithm

step t (lines 3 to 9). The falsification of a candidate leads to its removal from P_{cand} (line 6). After the loop ends, only property candidates are left in P_{cand} that are valid on T. To obtain high-quality properties it is necessary that the used trace has got a high functional coverage. Otherwise, a large number of properties are incorrectly inferred and do not hold on the design. There, the approach is independent of how the simulation data are obtained, e.g. by a conventional simulation using directed tests, or a constraint-based random simulation. After the first phase, the set P receives all valid basic properties (line 10). Additionally, each property is detailed by a set of extracted parameters.

The subsequent second phase (lines 11 to 22) examines the temporal dependencies between already found properties in P. At first, new candidates of complex properties are created and stored in the set P_{cand} (line 13). Valid properties are combined to property candidates using the requested temporal dependencies that are indicated by dedicated checker types in $C_{complex}$. The next step checks the validity of each candidate on the trace and removes it from the candidate set if it has been falsified (lines 14 to 20). Valid properties extend P (line 21). Repeating this procedure, temporal dependencies between already found and newly generated properties are examined. So, even more complex expressions are generated and checked on the simulation trace. The algorithm loops as long as further candidates could be proven. Finally, P contains all basic and complex properties that are valid on T.

2.2.3 Complexity and Termination

The complexity of the generation algorithm (see Figure 5.2) composes of the complexity of the first and the second generation phase.

During the first phase the simulation trace T is processed one time (line 3). So, the complexity is linear to the number of clock cycles t_{cyc} stored in the trace, i.e. $O(t_{cyc})$. At each time step t, all instantiated property candidates in P_{cand} are checked for validity. The concrete number of created candidates (line 2) is $|P_{cand}| = |C_{unary/bus}| \cdot |S| + |C_{binary}| \cdot |S|^2 + |C_{ternary}| \cdot |S|^3$. That means, the number of basic candidates increases cubically in the worst case, i.e. $|P_{cand}| = O(|S|^3)$ while a constant number of selected checker types in C_{basic} is assumed. So, $phase_1 = O(t_{cyc}) \cdot O(|S|^3) = O(t_{cyc} \cdot |S|^3)$.

The second phase runs as long as new candidates of complex properties could be validated at each of n iterations where the simulation trace is always completely processed. In that case, the complexity is $O(n \cdot t_{cyc})$ for both outer loops (line 12 and 14). Assuming that only binary temporal dependencies are examined, the number of complex property candidates is $|P_{cand}| = |C_{complex}| \cdot |P|^2$ (line 13), i.e. $|P_{cand}| = O(|P|^2)$ if a constant number of checker types is assumed. Hence, the complexity of the second phase is $phase_2 = O(n \cdot t_{cyc}) \cdot O(|P|^2) = O(n \cdot t_{cyc} \cdot |P|^2)$ resulting in a complexity of the overall algorithm: $O(phase_1) + O(phase_2) = O(max\{phase_1, phase_2\})$. Generally, the generation algorithm does not reach the upper bound as later demonstrated by the experiments starting on page 128.

The termination of the algorithm can be shown as follows. During the first phase each basic property candidate p of the finite set P_{cand} is checked at each time step t of the given simulation trace with t_{cyc} clock cycles. Assuming that the functions *create_basic_cand* (line 2) and $check_p$ (line 5) terminate, the first phase terminates, too. The same loop is part of the second phase (lines 14 to 20) where it is assumed that the functions *create_complex_cand* (line 13) and $check_p$ (line 16) terminate. To let the

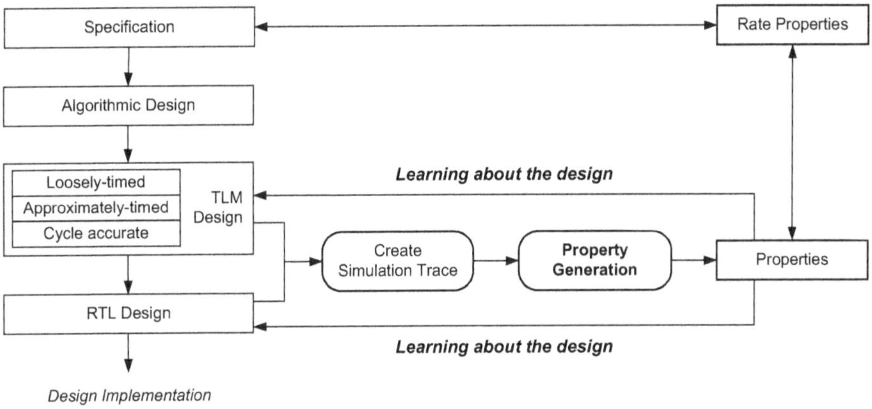

Figure 5.3: Property generation in the design flow from Figure 2.2

outer loop terminate (lines 12 to 22), two additional assumptions are made for the function *create_complex_cand* in line 13:

1. Only new property candidates are created at each iteration.
2. A complex property candidate is only admitted, if it characterizes a larger part of the simulation trace.

According to assumption 2, two cases are distinguished: First, the new candidate embraces one more signal. So, after a finite number of iterations all signals in *S* would be part of a possibly valid complex property. Second, the new candidate does not embrace a new signal but covers more clock cycles. Both cases lead to a point, where a new complex candidate cannot be created and the algorithm terminates.

2.3 Design Flow Integration

The integration of the property generation methodology while designing the hardware part of an SoC is shown in Figure 5.3. As soon as the system model or the RTL design can be simulated, a trace is created and used to hypothesize complex properties. These inferred properties abstract important design behavior. Moreover, the properties help the designer to learn new aspects about the simulated design. Since generated properties can represent incomplete or wrong behavior, due to an insufficient database, the designer has to check their correctness with respect to the design intent and the specification. This interactive but time-consuming process improves design understanding. Furthermore, it allows to detect inconsistencies in the specification, gaps in the test suite, or errors in the simulated design. If properties

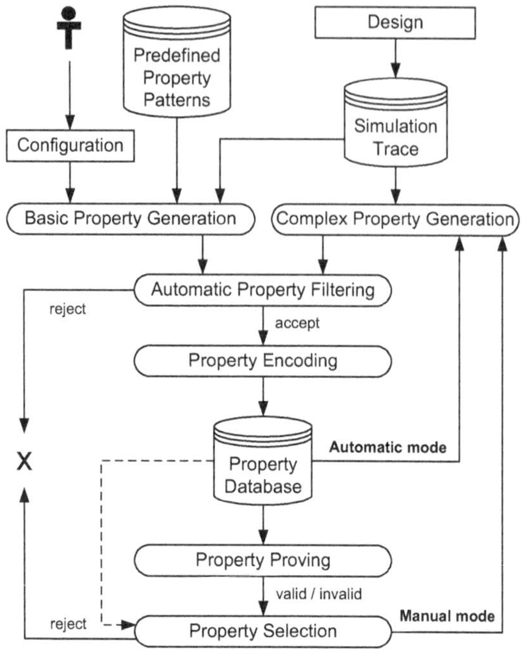

Figure 5.4: DIANOSIS property generation flow

turn out to be correct, they can be used as another kind of design specification. Properties can be used as a starting point for formal verification approaches by converting them into assertions or temporal logic expressions. This procedure reduces the effort of property writing and simultaneously increases the acceptance of formal techniques. Another valuable approach is given by a comparison between properties inferred over passing simulation runs and the ones found on failing runs. Here, missing or additionally detected properties can be a good indicator of defects.

3 DYNAMIC INVARIANT ANALYSIS ON SIMULATION TRACES

This section sketches the implementation of the proposed generation methodology. Based on the algorithm presented in Figure 5.2, the DIANOSIS tool was developed. Figure 5.4 depicts the DIANOSIS property generation flow. First of all, a simulation trace in the VCD format is created by simulating the desired design. Then, the generation process is configured that comprises choosing a set of checkers, defining reset and clock scheme, and optionally specifying a regular expression to restrict property generation to certain signals. During the first generation phase, basic properties are inferred (see Section 3.2 on page 118). Properties that have "survived" over the complete simulation trace are defined as valid properties. Subsequently, each generated property runs an automatic filter to filter out likely wrong design assumptions (see Section 3.3 on page 122). Accepted properties are encoded and added to the property database. The encoding phase defines new virtual signals which enable an equal handling of properties and signals (see Section 3.4 on page 123). Then, complex properties are iteratively inferred over the already found properties (see Section 3.5 on page 123). These complex properties are filtered, encoded, and added to the property database, again. In contrast to this automatic mode, the user manually selects the generated and optionally proven properties for the (next) combination phase running the manual mode (see Section 3.6 on page 126). If no new complex properties could be inferred the generation process ends and the user selects the correct properties out of the generated set. Before, the different generation phases are detailed, the general architecture of DIANOSIS is described.

3.1 General Architecture

DIANOSIS has a layered architecture that is depicted in Figure 5.5. Each of the three tiers *Front-end*, *Data Storage*, and *Back-end* communicates with the next tier using defined interfaces. This eases maintenance and facilitates extension with further functionality. The provided VCD simulation trace is read in by the *VCD Parser*. Using the VCD format, properties can be generated independently of the simulator tool, the abstraction level and the design language. Signal states are stored in the *VCD Dump* component. There, only signal changes are saved which limits memory consumption (see Section 3.7 on page 128). The design hierarchy is analyzed in parallel and stored in the *Design Hierarchy* component. With this, property generation can be limited to specific modules. The *Checker Controller* implements the generation algorithm from Figure 5.2. Each property type is validated by an appropriate checker. *Unary*, *Binary*, and *Ternary Checkers* analyze scalar signals whereas *Bus Checkers* process multi-bit signals. On request the *Checker Factory* creates a checker instance for each property candidate. If a property candidate is valid over the whole simulation trace, it is stored into the *Property Database*. Hence, it is available to be combined to more complex expressions in the next iteration. Finally, the resulting properties can be converted into different formats suitable for advanced verification tasks such as SVA, or PSL expressions.

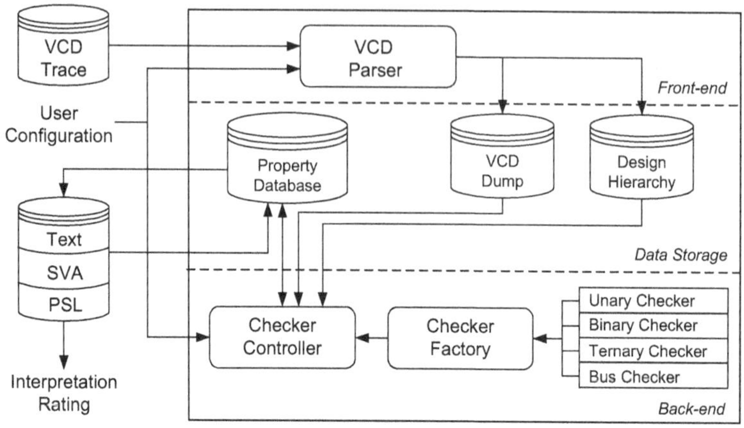

Figure 5.5: DIANOSIS general architecture

3.2 Basic Property Generation

To generate valuable properties, the availability of a large set of basic properties that can be combined is essential. Moreover, mainly typical design behavior shall be searched. Several libraries and different work [OVL], [DAC99], [Vera] encapsulate common design behavior, reusable verification IP or property specification patterns. Using these components the designer can quickly create complete verification environments. As a practical starting point, the OVL is selected as a template to search for typical behavior in designs. Currently, the library provides a set of 51 configurable assertion-based checkers. The functionality ranges from simple checks such as the test for one-hot coded buses to complex checkers that validate a user-specified arbitration scheme. Due to the generality of OVL in validating basic design behavior, a couple of assertion checkers is chosen. Therefore, proper search algorithms in terms of checkers are implemented in DIANOSIS. Beyond this, a number of user-defined checker types were realized, as well. Currently, DIANOSIS implements 16 basic property checkers having an arity of 1 – 3 signals. Table 5.1 gives a selection of typical representatives. There are further basic checkers imaginable that are not yet implemented. Such checkers could be for instance a state checker that records all states and transitions of a state variable, or some checkers to characterize data transmitted on a bus.

Table 5.1: Selection of basic property checkers

Checker	*Type*	*Checks for ...*
Const_Signal	Unary (User)	a constant signal
Bus_Trigger	Binary (User)	a scalar signal that always triggers the same value on a bus
OVL_Handshake	Binary (OVL)	signal pairs that follow the OVL handshake pattern
Req_N_Grant	Binary (User)	signal pairs where one signal changes the state whenever a second (trigger) signal remains active
Req_Grant_K	Binary (User)	a signal that triggers a second signal K-times
Req_Grant1_Grant2	Ternary (User)	a signal that triggers two other signals that may not be active together
Bus_Mutex	Bus (User)	specific values that are mutually exclusive on two busses
FIFO_Transfer	Bus (User)	values on two busses that are transferred from the source to the target bus following a FIFO behavior
OVL_Increment	Bus (OVL)	an incrementing counter
OVL_OneHot	Bus (OVL)	a one-hot coded bus
Shifter	Bus (User)	a shifting behavior on a bus

Definition 12. *Each property checker c with* $c \in C$*, regardless of checking basic or complex properties, is a 7-tuple FSM* $c = (\Sigma, \Lambda, Z, z_{init}, \delta, \lambda, F)$ *with*

- $\Sigma = (a_{s_1}[t], \ldots, a_{s_n}[t])$*: the input alphabet representing a tuple of signal activities of each participating signal* s_i *at time t with* $t_{rst} \leq t \leq t_{rst}$ *and* $n = 1$ *for unary and bus,* $n = 2$ *for binary, and* $n = 3$ *for ternary checker types,*
- Λ*: the output alphabet representing the outcome of the checker i.e. either a valid property candidate with a set of extracted property parameters or a falsified one,*
- *Z: the finite set of states representing the different analysis steps to validate the property candidate,*
- z_{init}*: the initial state* $z_{init} \in Z$ *in which each property candidate is assumed to be valid,*
- δ*: the state-transition function* $\delta : Z \times \Sigma \rightarrow Z$*,*

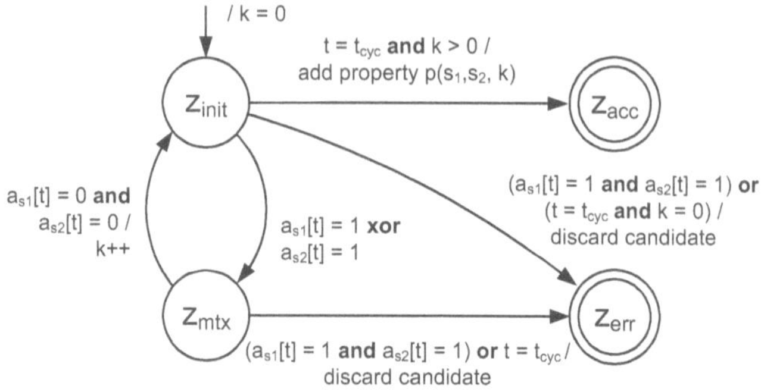

Figure 5.6: Finite-state machine of the mutual exclusion checker

- λ *: the output function* $\lambda: Z \times \Sigma \rightarrow \Lambda$*, and*
- $F = \{z_{err}, z_{acc}\}$ *: the set of final states* $F \subseteq Z$*. There,* z_{err} *denotes an error state identifying the falsification of the property candidate and* z_{acc} *denotes its acceptance.*

If the FSM reaches the error state z_{err}, the checked property candidate is deleted from the candidate set P_{cand}. The FSM accepts the candidate (state z_{acc}) only when the specified behavior has been seen at least once, the trace was read completely ($t = t_{cyc}$), and the FSM is in state z_{init}, again. Otherwise, the property could not be fully recognized and the candidate is discarded, as well.

Example 21. *Figure 5.6 depicts the checker that determines mutual exclusive activation of two signals* s_1, s_2 *in terms of their activities* $a_{s_1}[t]$ *and* $a_{s_2}[t]$ *at each time stamp t. If the mutex behavior is falsified, the FSM goes into the error state* z_{err} *and the property candidate is removed. If only one of the two signals is active, the checker has recognized a new mutual exclusion activity (state* z_{mtx}*) which does not end until both signals are inactive again (leaving state* z_{mtx}*). The variable k counts each recognition of a mutual exclusion activity. After the simulation trace is completely processed, the FSM checks whether at least one mutex behavior has been detected (k > 0) and if so, marks the property candidate as valid on the trace (state* z_{acc}*).*

Before basic properties are inferred, the simulation trace is read. There, DIANOSIS performs various actions:

- Add user-defined clock and reset signals to S_{EX}.
- Estimate signal polarity by using the signal activity heuristic from Equation (5.1).

- Check for constant signals and add them to the set of excluded signals S_{EX} with $S_{EX} \subset S$. Signals in S_{EX} are excluded from any property generation.
- Determine all scalar and multi-bit signals contained in the trace that matches the user-specified regular expression *rx* and add them to the set of matching signals S_M with $S_M \subseteq S$.
- Analyze the design hierarchy to generate properties module-wise. So, the generation process is tailored to related design parts which minimizes the probability to analyze unrelated signal pairs.

In the next step, over all signals in S_M basic property candidates are created over all selected checker types, i.e. accordant checkers are instantiated. The number of basic checkers can quickly become large where their number is of cubical order (see Section 2.2 on page 111). Hence, the creation of candidates is subject to several restrictions:

- Discard binary and ternary property candidates over combinations of a signal with itself.
- Exclude candidates over signals that are contained in S_{EX}.
- Treat only one combination of a symmetric property, i.e. a property where the signal order is irrelevant.

Furthermore, the given restrictions increase the likelihood of inferring reasonable properties. If the simulation trace contains a large number of signals, the generation process should be tailored to interesting signals using the provided regular expression *rx*. Here, especially the compliance of signal names to specific naming conventions could facilitate the choice of (related) signals. Static analysis, as described in Chapter 3, could ensure such naming conventions by means of coding standards.

During an update step the current activities of all signals participating in still valid property candidates are passed to the appropriate checker instances. To speed up the execution, a checker is only invoked when the observed signal values have changed. After that, the simulation time is shifted to the next value changes. As already mentioned, each checker type is individually implemented as a FSM. Depending on the complexity of the analyzed property the checker comprises up to several hundreds of lines of C++ code. Encapsulating each property checker into its own FSM allows us to switch on or off checks for particular properties very easily. Moreover, this procedure eases the handling of an arbitrary number of signals and property checkers.

The first generation phase terminates when no more candidates are left or the end of the simulation trace is reached.

3.3 Property Filtering

Each generated basic or complex property is automatically filtered. The user specifies a confidence level defining the minimum number of occurrences a property has to be seen on the trace. So, properties can be rejected that reflect only a random observation of the design behavior caused by an incomplete simulation trace. Hence, the generation process is sped up by further considering only the most promising properties. Finding the "right" confidence level could be an interactive process. Its value is important for the number of created complex property candidates, and thus the number of algorithm iterations (see Section 2.2 on page 111). A too loose confidence level allows too many properties, possibly describing random behavior, to reach the next combination phase. In contrast, a too strict level possibly does not leave any property for a further combination. Accepted properties are encoded (see Section 3.4 on page 123) and added to the property suite.

There are two ways to handle the encoded properties either using an *automatic mode* or a *manual mode*. Within the automatic mode, DIANOSIS runs without any user interaction until the algorithm terminates. Then, the inferred properties are optionally passed to an external proof engine where the result is presented to the designer for manual inspection the first and only time. To support different proof engines, DIANOSIS has to output the properties in the particular input language. The automatic mode is faster but needs more effort in reviewing a larger set of, potentially wrong, properties.

In the manual mode, the user selects only correct properties for the (next) combination phase which accelerates the generation process. The proof engine can be optionally used again to filter out wrong properties. In case a property cannot be proven, it is either a random observation or it only describes the expected behavior, incompletely. For illustration, assume that DIANOSIS infers the "Incrementer property" on a data bus (see Table 5.1). The counting behavior is created by the test bench writing incremented values onto the bus. Now, the proof engine would not be able to verify such a bus behavior. A further advantage of a proof engine is given by the reported counter example. Herewith, an incomplete property can be completed. For instance, it should be assumed that the maximum acknowledge cycle of a handshake is not detected correctly due to missing corner-case tests. The proof engine shows this by a counter-example which allows to fix the cycle number.

3.4 Property Encoding

After properties have been inferred and filtered, all occurrences of behavior corresponding to a certain property is composed in a single transaction. The goal of this composition is a similar, and thus simplified handling of found properties and given signals during subsequent combination iterations. A transaction describes the property activity in terms of the activity of each participating signal. During property generation a variable k counts how often the property behavior has been seen (e.g. variable k in Example 21). Each time a property is inferred from the trace, the start time of the property behavior and its end time are recorded. The interval $[t_s^k, t_e^k]$ relates to the kth occurrence of the property. This is used to define the activity $a_{p_i}[t]$ of a property p_i based on the activity of the considered signals s_i at each time stamp t:

$$a_{p_i}[t] = \begin{cases} 1 & \exists k: t_s^k \leq t \leq t_e^k \\ 0 & otherwise \end{cases} \tag{5.2}$$

In case of *unary* and *bus* properties the original signal activities are treated as a single transaction. As a result of the encoding phase m new activities $(a_{p_1}, \ldots, a_{p_m})$ for the found *binary* and *ternary* properties extend the simulation trace T. With this, the design state at time t is augmented: $v[t] = (a_{s_1}[t], \ldots, a_{s_n}[t], a_{p_1}[t], \ldots, a_{p_m}[t])$.

3.5 Complex Property Generation

Next, complex properties are combined over already found properties. Several experiments have shown that property combination can be also done using a simple string pattern matching over signal names, e.g. particular handshakes that always follow on each other. So, the actual combination algorithm is introduced by a preprocessing step that currently implements four different operations:

- *Handshake_Sequence*. If a handshake acknowledge signal requests a further handshake (see Table 5.1), both handshakes are connected. This operation is repeated until no new handshake can be connected to an existing sequence.
- *Bus_Trigger_Join*. Bus_Trigger properties (see Table 5.1) triggered by the same signal are composed into a joined Bus_Trigger property. Based on that trigger signal always the same values are seen on each bus.

- *Req_N_Grant_Join.* Req_N_Grant properties (see Table 5.1) having the same trigger signal are combined into a joined Req_N_Grant where the common request signal releases all triggered grants.
- *Handshake_Cluster.* All handshake properties having the same trigger event are subsumed into a common cluster. Such a cluster replaces the contained handshakes in the set of already found properties. So, without loose of expressiveness the temporal relations *Response* and *Equal_State* can be explored.

After the preprocessing step, the actual property combination starts. This process exploits so called "temporal dependencies". In fact, temporal dependencies reflect specific design knowledge. Exemplarily, this knowledge was obtained by intensive discussions with the SIMD designers (see Section 4.2 on page 131). Here, especially the examination of the found basic properties and their temporal relationships gave valuable hints for reasonable combination operations. Dwyer et al. [DAC99] present so called property patterns that can act as a second source for the definition of valuable temporal dependencies. To combine properties p_i and p_j to a new property p_{new}, the following dependencies are checked:

- *Mutual_Exclusion.* The properties p_i and p_j may not be active at the same time:

$$p_{new} \text{ is valid } \leftrightarrow \text{ for each } t_{rst} \le t \le t_{cyc} : \neg(a_{p_i}[t] \wedge a_{p_j}[t]) = 1 \tag{5.3}$$

- *Response.* Property p_j always follows property p_i:

$$\begin{gathered} p_{new} \text{ is valid } \leftrightarrow \text{ for each } t_{rst} \le t \le t_{cyc} - l : \\ a_{p_i}[t] = 1 \rightarrow a_{p_j}[t+l] = 1 \text{ with } l \ge 1 \wedge a_{p_i}[t-1] = 0 \end{gathered} \tag{5.4}$$

- *Equal_Activity.* The properties p_i and p_j are always active at the same time:

$$p_{new} \text{ is valid } \leftrightarrow \text{ for each } t_{rst} \le t \le t_{cyc} : a_{p_i}[t] \wedge a_{p_j}[t] = 1 \tag{5.5}$$

- *Equal_State.* Property p_i describes a multi-bit signal which has always the same state when property p_j becomes active:

$$\begin{gathered} p_{new} \text{ is valid } \leftrightarrow \text{ for each } t_{rst} \le t \le t_{cyc} : \\ a_{p_i}[t] = a_{p_i}[t^1_{p_j}] \wedge (a_{p_j}[t] = 1 \wedge a_{p_j}[t-1] = 0) \end{gathered} \tag{5.6}$$

$t^1_{p_j}$ *indicates the first time* p_j *becomes active*

In addition to the restrictions applied for the creation of basic candidates, property combination is further limited:

- Do not consider combinations that produce meaningless results, e.g. the check for a mutual exclusion between two counter signals.
- Create only new candidates. Hence, each iteration always generates more complex properties.
- A complex property candidate is only admitted if it characterizes a larger part of the simulation trace.

Composed property candidates are analyzed over the simulation trace in the same manner as basic properties. So, "surviving" candidates are filtered, encoded and inserted into the property database to analyze their temporal dependencies in the next iteration. The algorithm terminates if the current iteration does not find a new property.

In a postprocessing step, DIANOSIS subsumes all inferred properties into *property clusters* to improve readability. Starting with a handshake cluster, a property is added to that cluster if all its signals already affiliate to one of the handshakes contained in the cluster. Hence, behavior that belongs together is summarized in a cluster which helps the designer to easily interpret interrelated behavior.

Example 22. *Figure 5.7 depicts property generation in case of four signals* `req`*,* `ack`*,* `en`*, and* `rdy`*. In the first phase, two handshake properties are found between* `req` *and* `ack`*, and* `en` *and* `rdy` *(see Figure 5.7(a)). Then, the handshakes are encoded to describe the property activity (see Figure 5.7(b)). So, each (req, ack) pair is composed into a transaction where a transaction starts when the req-signal becomes active and ends when the corresponding ack-signal gets inactive again. To limit memory consumption, the property activity is characterized only by value changes at the particular time stamps. According to Equation (5.2), the following encoded sequence is reported:*

$$a_{p_1} = a_{req \rightarrow ack} = \begin{matrix} & a_{p_1}[0] & a_{p_1}[1] & a_{p_1}[4] & a_{p_1}[14] & a_{p_1}[18] & \ldots \\ (& 0, & 1, & 0, & 1, & 0, & \ldots) \end{matrix}$$

Last, both basic properties are checked to be mutually exclusive (see Figure 5.7(c)). Assuming that the mutex candidate "survives" over the simulation trace, DIANOSIS reports the SVA property expression: `not((req ##[2:3] ack) and (en ##[2:3] rdy))`[1].

1. To keep this, and all following, property examples as simple as possible, the found handshake properties are given as property sequences. In reality, DIANOSIS reports a full OVL_Handshake property.

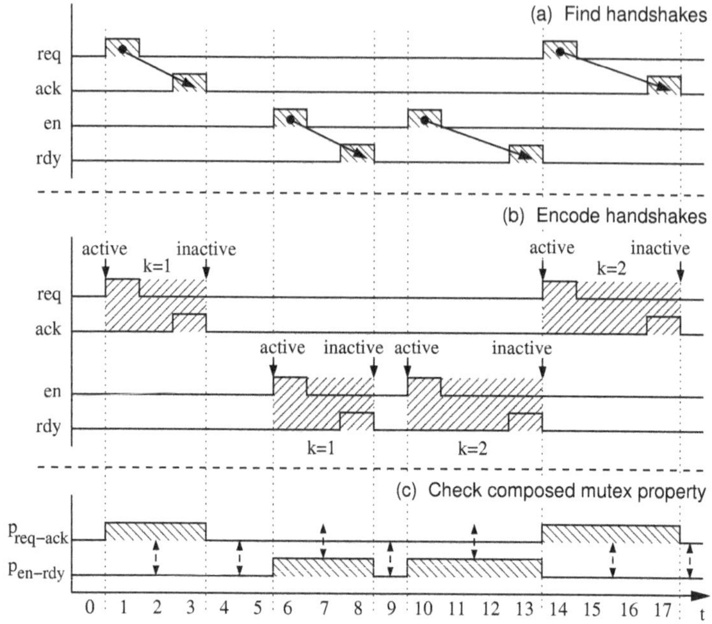

Figure 5.7: Property generation example

3.6 Property Selection

DIANOSIS offers various features to interpret generated properties and to support the selection of correct ones. A log file contains all falsified property candidates. This file helps the user to evaluate why an expected property could not be detected which assists debugging of the design. The following log sketches an example showing the property type, a short error description, the participating signals, and the falsification time for three disproved property candidates:

```
PROPERTY DESCRIPTION SIGNALS TIME
Req_N_Grant         [no grant on req]     en_o, ack_i     @66ns
Req_Rep_Grant       [request underrun]    req_sc, ack_i   @69ns
Req_Grant_Rep       [grant count = 1]     grd_o, ack_i    @69ns
```

In addition to the data describing each found property, its occurrence in the simulation trace is counted and stored as well. Thus, properties can be ranked to present the most frequent, and thus most likely ones first. Moreover, the most complex properties are reported at first. Hence, the designer can select properties that describe the design behavior more comprehensive than others. To present inferred properties in an easy to read representation, prop-

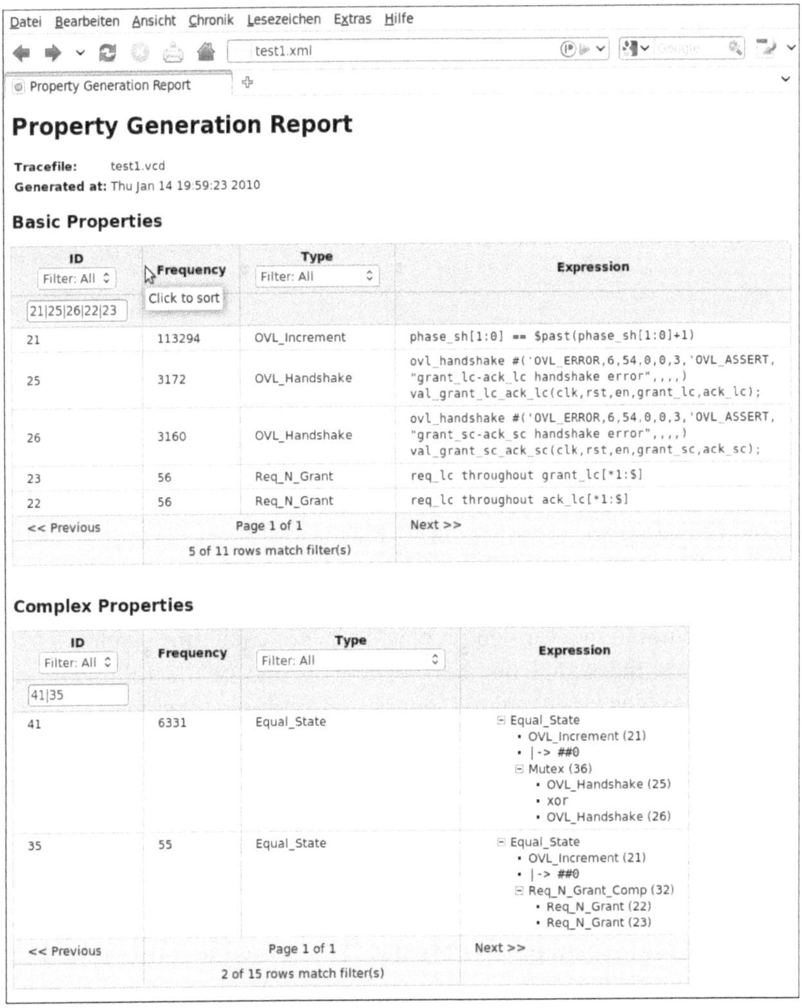

Figure 5.8: Example for a property generation report

erty descriptions were converted into an HTML report. An example is shown in Figure 5.8. Each table column can be sorted according to the contained data type. The ID and the Type column provide filters to restrict the output to specific properties as well as property types. The property itself is displayed as a compact and unambiguous SVA or OVL expression. For the sake of readability complex properties are presented as a tree that can be collapsed or expanded. Thus, uninteresting parts can be explicitly hidden and the designer gets a fast overview about the generation results.

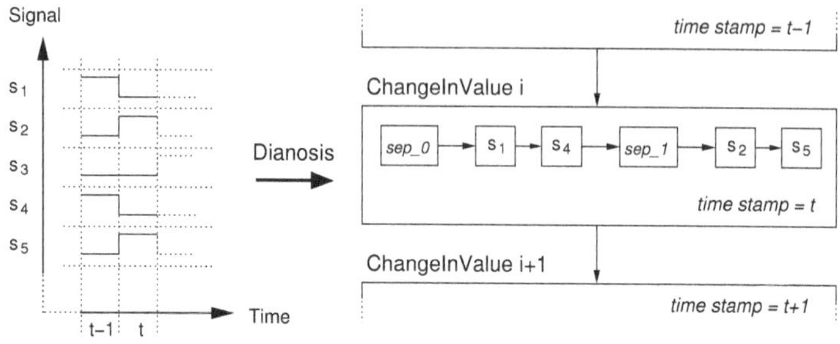

Figure 5.9: Efficient storage of signal values

3.7 Implementation Issues

To allow processing of industrial simulation traces that commonly consist of many hundred megabytes, the memory management of simulation data was optimized in DIANOSIS.

Due to efficiency reasons, DIANOSIS completely reads the trace data into the memory. There, it is sufficient to store only changes of signal values which is comparable to a run-length encoding. For each time stamp t a single-linked list holds the signal numbers sorted by one of the following signal states '0', '1', 'x', or 'z'. A unique separator introduces all signals with a particular signal state at t. Figure 5.9 sketches the implementation which allows to efficiently store a large amount of simulation data. At the considered time stamp t the signals s_1 and s_4 change to '0' while s_2 and s_5 change to '1'.

To further reduce memory consumption, the single-linked list holding the signal values was adapted to the particular target architecture. An identifier (ID) for a signal is stored as a short integer (16 bit) which corresponds to 65,536 signals and separators. If only one ID is stored in a list node, 48 padding bits would be inserted in case the application is compiled at a 64-bit architecture. Hence, the correct addressing of the next list node is ensured (see Figure 5.10(a)). However, the padding bits in each list node have room for three more IDs which yields an optimal memory utilization (see Figure 5.10(b)).

4 EXPERIMENTAL RESULTS

The following section presents results obtained by DIANOSIS on different designs. In the first part, DIANOSIS is applied to the SIMD data transfer example (see Section 3.4 on page 28). This example illustrates how property generation could help the designer to improve design understanding as well as to

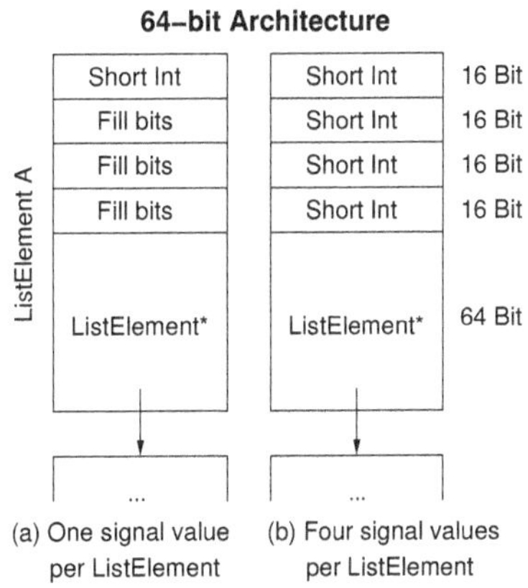

Figure 5.10: Optimization of the workload

isolate a hard to find defect. The second part documents promising results while applying property generation on different industrial RTL hardware designs.

4.1 SIMD Data Transfer Example Continued

To enable property generation on the SIMD data transfer example, the design is extended by a *monitor* component. This component observes the communication behavior and writes particular events in terms of transaction activities into a VCD file. Since a correct communication between the processor cache and the shared memory is a crucial design part, the transaction activity is analyzed by DIANOSIS. There, the missing cycle-accurate timing in the example only allows to infer the general functional behavior.

To keep the SIMD example simple, it models only the communication between the processor unit and the shared memory. Specific load and store instructions initiate a data transfer while all other SIMD instructions are not evaluated especially. Due to this simplified behavior, random, not necessarily meaningful but correct programs can be generated using the instruction set of the SIMD processor. The randomization process reproduces typical programs reflecting a similar distribution of instruction usage as in real SIMD programs. Then, these randomly created programs are simulated on the SIMD example design to generate many different simulation traces. Subsequently, these traces are analyzed. 1,009 random programs were generated and simulated. The resulting 1,009 simulation traces are fed into DIANOSIS to generate 1,009 property sets. In the following, two use cases are discussed: improve *design understanding* by inferring correct functional design behavior and detect *abnormal behavior* that could indicate a defect.

4.1.1 Improve Design Understanding

DIANOSIS infers 10 correct properties – 7 basic and 3 complex properties over all 1,009 property sets. They describe the general communication behavior between the processor unit and the shared memory. Here, one complex property is of special interest. It is composed of the two basic handshake properties:

```
arb.ld_req ##[1:$] arb.ld_ack
arb.st_req ##[1:$] arb.st_ack
```

The handshakes represent the transfer of a single data word from the shared memory to the processor cache and vice versa. During property combination, DIANOSIS detects a mutual exclusion between both handshakes and creates the following complex property:

```
not( (arb.ld_req ##[1:$] arb.ld_ack) and
     (arb.st_req ##[1:$] arb.st_ack) )
```

This property states that load and store transfers never happen at the same time. This finding proves the proper working of the arbiter in the load/store controller unit. Since this functional behavior is an important design requirement, it must be also followed in the final hardware design. So, the above complex property is formulated as an assertion and proved during following development steps, continuously. When the abstraction level is changed from ESL to RTL, the generated ESL assertion and the inferred property at RTL are checked for equality. A mapping mechanism between the ESL design and the proper RTL components enables the equality check. Therefore, the particular SIMD hardware block has been also analyzed by DIANOSIS (see Section 4.2 on page 131) where a similar complex property was detected. Again, this property describes the arbitration scheme of the arbiter hardware component which is part of the load/store controller unit. Due to the cycle-accurate description at RTL, the correct timing is additionally inferred:

```
ESL: not((arb.ld_req ##[1:$] arb.ld_ack) and
         (arb.st_req ##[1:$] arb.st_ack) )

RTL: not((grant_lc ##[2:18] ack_lc) and
         (grant_sc ##[2:18] ack_sc))
```

The two generated, functionally equivalent, properties show that the underlying design requirement has been kept during development. Summarized, property generation improves design understanding and helps to learn important design behavior that must remain valid during the design stages.

4.1.2 Detect Abnormal Behavior

A detailed analysis of the generated property sets shows a difference from the *usual* number of inferred properties in case of 13 sets (1.29 % of all sets). A comparison with an assumed correct property set indicates the absence of the basic handshake property `core.ld_req ##[1:$] core.st_req` and two complex properties which had recombined this property. So, an anomaly is found possibly indicating a defect. To analyze the abnormal behavior, the generated log files could help to explore why the particular handshake was falsified (see Section 3.6 on page 126). All 13 files, which have logged the abnormal generation process, contain a similar entry for the falsified handshake:

```
OVL_HANDSHAKE [ack before req] core.ld_req,core.st_req  @<time>
```

This log message indicates that the write data transfer does not start *after* the read data transfer at the specified time. In a waveform viewer the parallel beginning of read and write transfers can be traced. Counting the transfer blocks in the waveform viewer allows to find the matching source code block that initiates the erroneous data transfer. To isolate the error, the particular block and its instructions are further examined. A manual comparison between all erroneous transfer blocks in the 13 files shows a sharing of a certain instruction sequence. If a combined *multiply–subtraction instruction* is followed by a *maximum calculation* directly before data are written back into memory, write and read transfers are started simultaneously. A look into the model description indicates an erroneous synchronization of the data transfer in case of that specific and rare instruction sequence. Instead to the tedious and manual comparison procedure, in Chapter 6 an approach is presented that detects the failure-inducing instruction sequence, automatically.

Investigating abnormal design behavior has denoted a defect that would be detected otherwise only with difficulty especially because of its rare appearance. So, the use of property generation helps the designer to systematically explore anomalies and to process erroneous designs faster.

4.2 Industrial Hardware Designs

To evaluate the introduced approach in an industrial context, DIANOSIS is applied to several blocks of different RTL hardware designs: an 8-bit RISC CPU controlling the traffic lights on a street crossing [Syn] and three real-world industrial hardware designs, i.e. a SIMD multiprocessor design (SIMD MP), a SATA FIFO interface and a DRAM controller interface. Table 5.2 summarizes the design characteristics, i.e. the number of studied blocks with their gate count, the signals per block used to generate properties,

Table 5.2: Test bench characteristics

Test bench	*Gates*	*Blocks*	*Signals*	*Max. cycles*	*Analyzed traces*
Traffic light control	8,706	1	17[a]	5,006	1
SIMD MP					
Cache arbiter	479	4	48	113,298	32
Address generator	5,876	3	44	990,438	26
Memory read IF	5,260	1	22	990,438	7
Cache controller	7,458	1	13	113,298	9
SATA FIFO	7,466	1	93[a]	13,763,441	1
DRAM controller	1,020k	1	176[a,b]	162,912	2

a. Exclude debug/test signals from analysis

b. Restrict trace analysis to signals of one data channel only

the number of clock cycles in the input trace for DIANOSIS, and the number of analyzed simulation traces. The trace lengths vary from a few KB to approximately 1 GB.

The high number of analyzed simulation traces in the SIMD design corresponds to an available regression test suite that runs various test programs on the processor. The traffic light controller represents a small design where only directed tests are part of the design package. In case of the SATA FIFO, a detailed test bench was provided by the designers that runs many different test scenarios in a single run. The verification of the DRAM controller was at an early design phase. So, only two traces could be analyzed. Since the external design behavior is of interest, it is sufficient to dump only the particular interfaces at the top-level blocks. In general, all signals which are contained in the provided VCD trace files are automatically added to the generation process. The only exceptions are debug/test signals, and signals where the same functional behavior can be observed, e.g. in case of the equivalent data channels of the DRAM controller.

4.2.1 Generated Properties

The results of the basic property generation phase are shown in Table 5.3. DIANOSIS generates a relatively small but relevant set of properties in a comparatively short time. There, the numbers in brackets (second column) are properties describing constant signals or properties with a count below the specified confidence level of "2". All other properties were cross checked by

Table 5.3: Found basic properties

Test bench	*Properties*	*Correct*	*Incomplete*	*Wrong*	*Analysis time*[a]
Traffic light control	7 (+5)	6	0	1	0.39 s
SIMD MP					
Cache arbiter	38	33	2	3	25 s
Address generator	30 (+6)	14	0	16	85.7 s
Memory read IF	28	22	6	0	51.5 s
Cache controller	14 (+1)	12	0	2	6 s
SATA FIFO	39 (+19)	29	8	2	57 m
DRAM controller	23 (+404)	3	9	11	42 m

a. *Test system:* AMD Opteron™ 248 processor @2200 MHz, 3 GB RAM

the component designers and classified into three categories: *correct*, *incomplete*, and *wrong* (columns 3–5). A correct property reflects the implemented functionality, completely and correctly. In contrast, an incomplete property only partially characterizes the expected design behavior but correctly expresses the general functionality. Finally, a wrong property results from a random observation.

One set of incomplete properties describes necessary but not sufficient conditions. So, the inferred handshake `rb_empty ##[1:$] new_instr` qualifies only a necessity where the buffer must be empty to process a new instruction. Contrary, an empty buffer does not necessarily yield a new instruction. The other set of incomplete properties results from a wrong correlation of signal activities in case of Req_Grant1_Grant2 handshakes (see Table 5.1). One of the grant-signals is triggered only sporadically by the request-signal but the cause is not the current request at time t but some previous one at time $t-n$. The trace analysis cannot figure out such semantical dependencies. Figure 5.11 illustrates an example where the property `req ##[2:3] (grant1 or ##1 grant2)` is wrongly inferred. Instead, the `grant2` signal is always triggered by the previous `req` signal.

A directed test bench usually covers only a small part of the functionality. So, a large amount of incomplete properties in the SIMD design were produced at the beginning. However, these properties could be ruled out using traces obtained from a regression test suite. So, 39 unbounded properties were identified and could be declared as correct, because the clock cycles between cause and effect differ between the various traces. For instance, the formerly bounded handshake `enable_cmd ##[4:58242] done_cmd` was

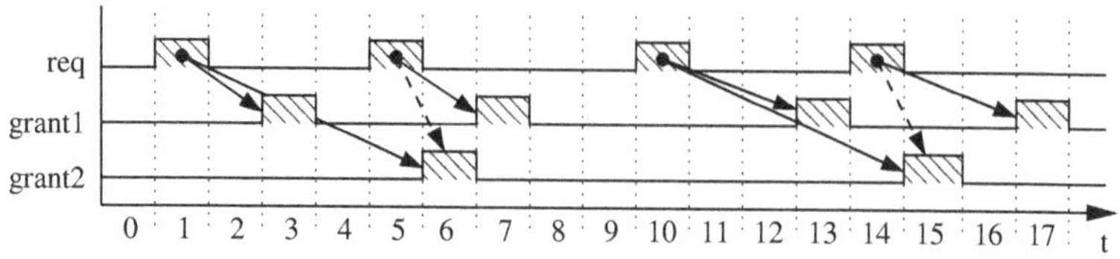

Figure 5.11: Incompletely inferred Req_Grant1_Grant2 property

detected as unbounded and DIANOSIS reports `enable_cmd ##[4:$] done_cmd`, instead. This corrected handshake states the different processing times of a command. The same observation of bounded and unbounded properties was made for the DRAM controller.

The multitude of wrong properties refers to buses. Especially the Bus_Mutex and the Bus_Trigger checkers (see Table 5.1) validate a lot of property candidates that describe only random observations. Hence, these property types should be used with care. Further random observations produce several wrong properties especially in the SIMD and the DRAM controller design. Here, incrementers and shifters have been identified as cause. The recognized behavior could be traced back to regular signal input stimuli which were wrongly or incidentally generated by the test bench. However, these properties only reach a low confidence of 2–3 occurrences on the simulation traces which facilitates their rejection.

Usually, incomplete or wrong properties indicate an insufficient simulation trace unveiling a bad test bench coverage. By this, these properties supply valuable information to improve the test suite. The experiments show that improvements concern among other things insufficient corner-case tests as well as inadequate randomization of input stimuli in case of directed tests.

Table 5.4 summarizes the results of the property combination phase. This phase requires only a small fraction of the time needed to infer basic properties. This fact is justified by the fact that only a small amount of complex property candidates is hypothesized over the simulation trace. The values in brackets (second column) are the number of found property and handshake clusters. Clusters facilitate property interpretation and readability, and thus are not dedicated as *real* properties. Note that only a few wrong basic properties are further combined which prevents the generation of wrong complex properties. For the DRAM controller no correct complex property is inferred. However, the three incomplete properties could be ruled out as correct if some more randomized simulation traces would be available. They describe some initialization sequences for data transfers with a fixed but unexpected timing.

Comparing the quality of inferred with hand-written properties is difficult. Some properties show a kind of cause-and-effect chain describing the interac-

Table 5.4: Found complex properties

Test bench	*Properties*	*Correct*	*Incomplete*	*Wrong*	*Analysis time*[a]
Traffic light control	1 (+2)	1	0	0	0.03 s
SIMD MP					
Cache arbiter	15 (+18)	13	2	0	4.4 s
Address generator	5 (+7)	4	0	1	12.1 s
Memory read IF	4 (+5)	4	0	0	4.6 s
Cache controller	1 (+7)	1	0	0	2.7 s
SATA FIFO	18 (+6)	12	6	0	7 m
DRAM controller	8 (+1)	0	3	5	18.6 s

a. *Test system:* AMD Opteron™ 248 processor @2200 MHz, 3 GB RAM

tion between multiple components that would not be explicitly written by a designer in this way. A formal specification for some blocks of the SIMD design, however, contains some of the generated complex properties. This observation and the designer feedback suggest that DIANOSIS partially generates complementary properties compared to those written by a designer.

4.2.2 Generation Statistics

Figure 5.12 exemplarily illustrates the number of valid binary property candidates over simulation time while analyzing the SATA FIFO design. It shows that after scanning 28.3 μs of the simulation trace, which corresponds to 0.01% of the complete simulation time, already 82.5% of the candidates were falsified. The step-like appearance of the graph can be traced back to single signals that are part of a couple of candidates. So, a change of these signals at a specific time invalidates many candidates at once. Figure 5.13 shows the number of valid candidates while combining properties for the SATA FIFO design. The combination phase consists of two iterations. The first one confirms 18 candidates. The second iteration finishes at 52.94 ms when the last remaining candidates are falsified.

Table 5.5 documents some statistics about inferred properties, more precisely the correct, incomplete, and constant ones. Columns 2–6 show the distribution between the property types, i.e. unary, binary, ternary, bus, and complex properties. Column 7 mentions the complexity depth that means the number of iterations used to generate complex properties. The property reuse rate (column 8) indicates how many properties are reused within more com-

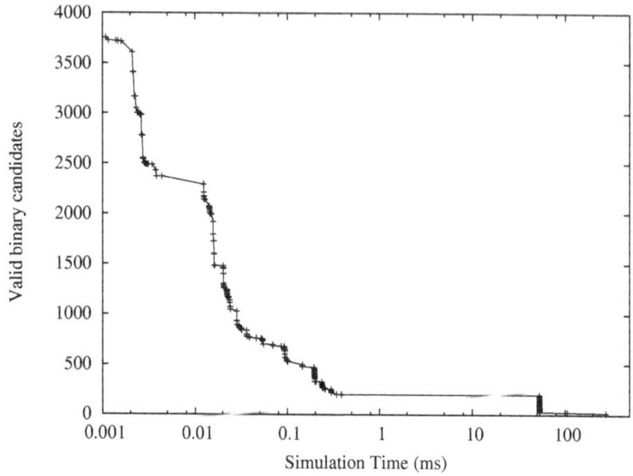

Figure 5.12: Valid binary property candidates for the SATA FIFO

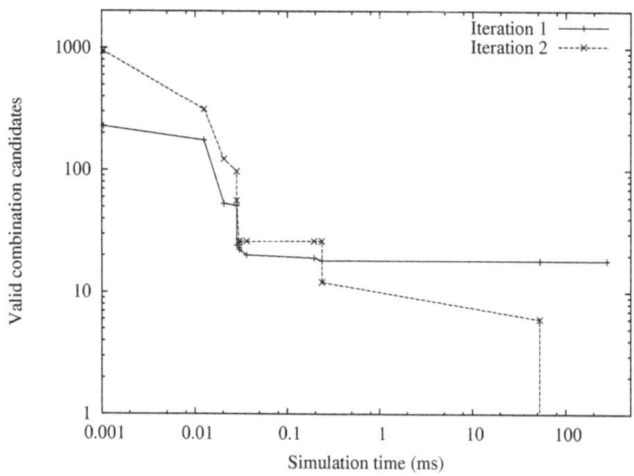

Figure 5.13: Valid complex property candidates for the SATA FIFO

plex ones. Finally, the last column (column 9) depicts the rate of all signals in the analyzed design block that participate in properties. The traffic light controller design has a high percentage of reported constant signals (the only unary checker). Constant signals usually indicate incomplete test benches that do not trigger all possible design behavior. Thus, a high value is a good indicator to make some effort to improve the test suite. An even higher percentage of constant signals is found for the DRAM controller. This is expected because the design was under construction at the analysis time. Thus, the test bench did not yet reach a very high functional coverage. The table also points out that binary properties are normally found most frequently where complex

Table 5.5: Statistical evaluation

Test bench	*Distribution of correct and incomplete properties + constant signals (%)*					*Complexity depth*	*Property reuse rate in %*	*Signal coverage in %*
	unary	*bin.*	*tern.*	*bus*	*comp.*			
Traffic light control	5 (41.7)	4 (33.3)	0	2 (16.7)	1 (8.3)	1	33.3	47.1
SIMD MP								
Cache arbiter	0	31 (62.0)	3 (6.0)	1 (2.0)	15 (30.0)	3	60.0	80.8
Address generator	6 (25.0)	14 (58.3)	0	0	4 (16.7)	2	83.3	56.7
Memory read IF	0	13 (40.6)	15 (46.9)	0	4 (12.5)	2	42.9	68.2
Cache controller	0	9 (69.2)	2 (15.4)	1 (7.7)	1 (7.7)	1	25.0	84.6
SATA FIFO	19 (25.7)	14 (18.9)	23 (31.1)	0	18 (24.3)	1	48.6	44.1
DRAM controller	68 (81.9)	12 (14.5)	0	0	3 (3.6)	1	58.3	47.2

properties reach up to 30%. This indicates the usability of the approach combining simpler properties to more complex ones. The fast termination of the generation algorithm is documented by a small number of iterations. An interesting value is the reuse rate. Especially, in the SIMD design up to 83.3% of properties are reused. This shows a close correlation between the signals in the particular block resulting in meaningful statements about the design behavior (see Section 4.2 on page 131). Moreover, the reuse rate can be a good indicator to explore the most significant properties first. The last column presents the number of signals participating in inferred properties. This value can be used by the designer to estimate the generation coverage that means how well the design behavior is covered by particular properties. Otherwise, the coverage depends on the number and types of implemented property checkers. Furthermore, it indicates how good the searched behavior matches the design behavior.

4.2.3 Case Study: Traffic Light Control

The traffic light control design is a freely available simple 8-bit microprocessor design that is binary code compatible with the Microchip 16C57. In [Syn] the design is extended by an expansion circuit that switches the traffic light at a street junction between a highway and a country road. A simulation trace describes the interface behavior of the expansion design. The two signals `hwy` and `cntry` control the traffic lights at highway and country road: 2 – *green*, 1 – *yellow*, and 0 – *red*. DIANOSIS reports a single interesting complex property (see Table 5.4):

```
Xtraffic && cntry == 0 && hwy == 2 ##[900:21900] hwy = 1
##[14000:49000] cntry = 2
```

This property is the result of a Bus_Trigger_Join combination (see Section 3.5 on page 123) of two Bus_Trigger properties. There, the `Xtraffic` signal triggers two different events:

```
P1  Xtraffic && cntry == 0 ##[35900:49900] cntry = 2
P2  Xtraffic && hwy == 2 ##[900:21900] hwy = 1
```

Normally, the highway has a green signal while the country road has red light. The activation signal `Xtraffic` indicates a car reaching the junction on the country road. This initial assumption is depicted by the left expressions of properties P1 and P2. Subsequently, the traffic on the highway is stopped by a yellow light followed by a red light (property P2). The country road is released by the green light (property P1). Combining both basic properties into a complex expression shows the correct temporal order and dependency of the events. Both events are released by the same trigger where the highway road switches to yellow *before* the country road gets the green light. The inferred complex property represents a more compact description of observed design behavior, and thus helps the designer to understand the design more quickly.

The other correct basic properties allow a further validation of the circuit behavior. Two Shifter properties (see Table 5.1) describe the proper switching sequence of both traffic lights. A partial mutual exclusion behavior between `hwy` and `cntry` bus signals (Bus_Mutex checker, see Table 5.1) denotes an important security issue. It indicates that except for the red light the traffic light control for both driving directions may not have the same state at the same time. The last property describes a FIFO-like data transfer (FIFO_Transfer, see Table 5.1) from the `hwy` control signal to the `cntry` signal which means that each driving direction is always switched in the same manner. Summarized, each of the generated properties helps the designer to understand abstract and important design behavior of the implemented system.

4.2.4 Case Study: SIMD MP Design

A number of interesting complex properties is found in the three industrial designs. Exemplarily, the SIMD MP design shall be chosen to illustrate the obtained results. The communication inside a SIMD processor unit is often synchronized by handshake signaling. So, amongst others four simple handshakes (basic properties) are reported by DIANOSIS:

```
finish_lc ##[15:$] ack_fin_lc
rb_empty ##[1:$] new_instr
req_lc ##[1:$] finish_lc
new_instr ##[1:$] req_lc
```

Between these handshakes, the *Response* dependency (see Section 3.5 on page 123) detects a temporal order and infers the following complex property:

```
rb_empty ##[1:$] new_instr ##[1:$] req_lc ##[1:$]
finish_lc ##[15:$] ack_fin_lc
```

This property is generated using the *Handshake_Sequence* operation in the preprocessing step of the complex property generation phase. It depicts the procedure to load data from the shared RAM into the processor local cache. A buffer decouples the shared RAM and the cache. If the buffer is empty (signal `rb_empty`), a new instruction may be loaded, which is indicated by the `new_instr` signal. Subsequently, available buffer data is written into the cache which is requested by the `req_lc` signal. Then, the `finish_lc` signal denotes that all data is read from the shared RAM into the buffer. Finally, the `ack_fin_lc` signal reports that all data is available in the local cache ready for processing. Thus, the property describes the complex data transfer behavior. It facilitates design understanding and allows a validation of design correctness at a higher functional level.

Another interesting property is composed of three basic properties:

```
grant_lc ##[2:18] ack_lc
grant_sc ##[2:18] ack_sc
phase_sh = $past(phase_sh+1)
```

The first handshake synchronizes a load operation from the global memory to the processor local cache while the second handshake controls a store operation from the cache. Additionally, an incrementer is reported. During the first combination iteration the *Mutual_Exclusion* dependency check finds a mutex between both handshakes:

```
not((grant_lc ##[2:18] ack_lc) and (grant_sc ##[2:18] ack_sc))
```

This behavior denotes the arbitration scheme of the arbiter and confirms its correct implementation. As result of the second iteration the *Equal_State* check detects a constant value of the `phase_sh` signal whenever a load or a store operation occurs:

```
not((grant_lc ##[2:18] ack_lc) and (grant_sc ##[2:18] ack_sc))
##0 phase_sh = 3
```

This complex property approves the correct synchronization of load/store accesses which is an important design requirement. Hence, this complex property helps the designer to understand and validate the abstract system behavior.

5 SUMMARY AND FUTURE WORK

In this chapter a new and promising debugging technique has been proposed. It enables the designer to learn certain new aspects about a design. Therefore, the findings of concrete simulation runs are summarized to general abstractions that hold for all runs. The proposed methodology automatically generates complex properties from a given simulation trace in an iterative manner. Using the standard VCD format for simulation data, the approach is independent of the simulator tool, the abstraction level of the analyzed design, and the design language. So, ESL as well as RTL designs can be processed. Starting from predefined basic properties, more complex ones are iteratively composed and checked over the trace. So, each iteration further abstracts the design behavior.

Experimental results, using the prototype tool DIANOSIS, have been shown that property generation is a valuable approach to enhance test bench coverage, to detect errors in the design or the test environment, and to identify holes or weaknesses in the specification. Properties were generated in a relatively short time ranging from the fraction of a second (0.42 s) to several minutes (57 m). Depending on the used simulation trace, 44.1– 84.6% of analyzed signals participate in inferred properties. These values indicate a good generation coverage. Correct properties can be used for arbitrary verification tasks and lower the barrier to successfully apply formal verification techniques. The interactive process of reviewing and discussing generated properties helps the designer to get a new and changed insight into the functional design behavior. The process of comparing property sets, generated from failing and passing simulation runs, documents how anomalies can be used while debugging a design. Summarized, property generation improves design understanding and supports debugging.

Future work could include the implementation of new basic property checkers that allow to infer further new properties. Moreover, the methodology could be combined with an approach that searches formerly unknown properties.

Chapter 6

Isolating Failure Causes

The so far introduced deductive, observational, and inductive techniques already facilitate and accelerate debugging of system models. However, the automatic determination of actual failure causes is not achieved by these techniques. This chapter proposes an *experimentation technique* that uses a series of experiments in terms of multiple controlled simulation runs of the fully integrated system. There, this debugging technique aims at an automatic and systematic isolation of failure causes (see Figure 6.1).

Isolating the cause of a failure requires to search for cause–effect relationships. Zeller [Zel05] defines this relationship as follows: "A *cause* is an event preceding another event without which the event in question (the *effect*) would not have occurred". Mapped to the debugging context, we could say: A defect causes a certain failure if the removal of this defect would eliminate the failure, as well. The idea is to narrow down the failure-inducing actual defect by comparing two program runs, one run where the effect (the failure) occurs and another program run where the effect does not occur.

The first part of the chapter gives a general overview about failure causes, their search, and their isolation. Then, a methodology for an automatic isolation of actual causes in SystemC design descriptions is introduced.

In the second part of the chapter, the ISOC tool (ISolation Of Failure Causes) is introduced. ISOC automatically narrows down the failure-inducing cause in SystemC designs using the delta debugging algorithm as proposed by Zeller and Hildebrandt [ZH02]. The SystemC scheduler is extended by a record and replay facility to deterministically rerun, formerly recorded, simulation runs. The algorithm is used exemplarily to isolate the minimal difference between two process schedules, one representing a passing (successful) simulation run, and the other resulting in a failing (erroneous) run. A second experiment uses delta debugging to detect the actual failure cause in simulation input data.

1 EXPERIMENTATION TECHNIQUES IN A NUTSHELL

First, this section summarizes some basics of experimentation techniques as introduced by Zeller [Zel05]. Next, the related work section discusses

F. Rogin and R. Drechsler, *Debugging at the Electronic System Level*,
DOI 10.1007/978-90-481-9255-7_6, © Springer Science+Business Media B.V. 2010

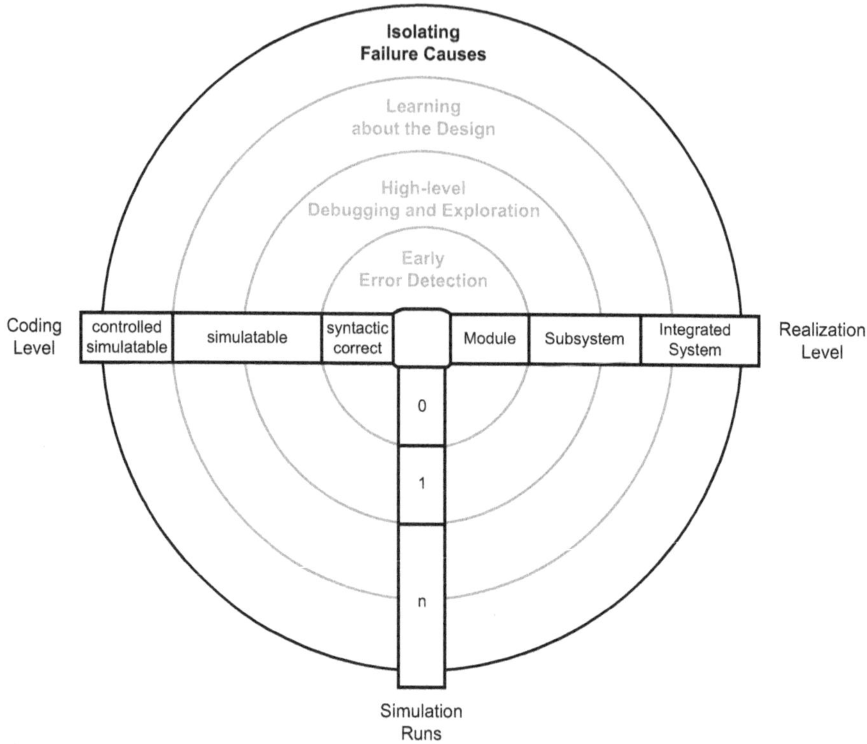

Figure 6.1: Automatic isolation of failure causes in system designs

several approaches that allow to detect typical failure causes in parallel programming languages, e.g. deadlocks, livelocks, or data races.

1.1 Overview

The experiment is a popular instrument in science that is used to systematically reject or confirm a hypothesis. On success, it shows the causality between a *cause* and the observed *effect*. A usual procedure is to set up a failing world, where the effect occurs, and an alternate passing world, where the effect does not occur. The experiment tries to find the difference between both worlds which represents the actual failure cause. In contrast to reality that does not allow to repeat the history to explore an alternative course of events, computers facilitate the deterministic replay of an experiment over and over again. So, the alternate world can be explored to check whether cause and effect relate to each other.

One of the most important problems of error search is the identification of the *real cause* for a failure. So, someone could think that the computer technology is the overall failure cause because without computers failing programs could not exist. To eliminate such obviously trivial alternatives, Zeller proposes the concept of the *closest possible world* [Zel05]. Here, the failing and the passing world shall only differ by means of the actual cause. This principle goes back to *Ockham's Razor*, which states that if two theories equally explain the same fact, the simpler theory should be picked. Relating to the above example, the unavailability of computers is far away from reality. On the other hand, there is no need, and sometimes no possibility, for a completely correct world but it should be possible to find a closest possible world where the failure does not occur. This difference spawns the search space to detect the actual cause. Concerning computer programs, the actual failure cause can be traced back to different sources such as the program input, the program state, or the program code. The iterative comparison of the closest possible world and the alternative world isolates the actual cause–effect relationship. An extracted failure cause does not necessarily point to the actual bug but gives valuable information where to start debugging. Here, the particular found cause could suggest a fix.

1.2 Related Work

Delta debugging was originally proposed by Zeller and Hildebrandt [ZH02]. They used the algorithm to isolate failure causes in the input of programs, in process schedules of parallel applications, or program code changes. There, a suitable test function and a decomposition strategy between a passing and a failing test case is used to narrow down the difference between both test cases. Misherghi and Su [MS06] suggest an improvement over the general delta debugging algorithm and call it *Hierarchical Delta Debugging*. Their approach considers the semantics and structure of data to early prune irrelevant parts of the input and to create simpler outputs. Hence, the search space is limited, and thus fewer test cases are needed to isolate the minimal difference between a passing and a failing test case. As result, more complex problems can be handled compared to the original delta debugging algorithm. Both works apply delta debugging in the software domain, while our approach extends the application field to support the debugging of system designs. To our best knowledge, the presented approach is the first work that uses this debugging technique in the SystemC context.

Delta debugging is able to isolate arbitrary failure causes. However, there are specialized approaches to detect or to prevent hard to find errors in parallel programs. Such problems are often caused by a violation of communication constraints or the erroneous synchronization between parallel components.

Cheung et al. [CS+06] present an approach that dynamically monitors a SystemC simulation and reports a deadlock once it has occurred. A dynamic synchronization dependency graph is created for this purpose. The vertices represent process dependencies and the directed edges denote synchronization dependencies. During simulation vertices and edges are dynamically added and deleted whenever a process suspends its execution. A recurring loop detection algorithm searches for cyclic dependencies afterwards. The approach has to know all processes that can notify a particular event for what static analysis techniques are used. Due to the nature of static analysis, notifying processes are approximated conservatively, i.e. the real number is typically smaller. A similar deadlock detection approach is proposed for the Metropolis environment [CD+05]. Metropolis is a complete system level design framework with its own model of computation. Both approaches dynamically report a deadlock once it has occurred. This could indicate the failure situation but does not suggest a fix. In contrast, delta debugging reports a minimal difference between a passing and a failing simulation run in a post-mortem analysis. Moreover, due to the generality of delta debugging, the analysis is not limited to specific language constructs used for synchronization as in [CS+06] or [CD+05].

A number of tools, e.g. [SB+97], [PS03], [YRC05], instrument programs to detect race conditions in software programs. Instrumentation allows to evaluate the locking discipline on shared variables. Generally, two approaches are distinguished. *Lockset-based tools* associate a lock candidate set to record all locks which are used to protect a shared location. When a thread accesses a shared resource, the intersection between the thread locks and the particular location locks may not be empty. A more sophisticated analysis performs a so called *happens-before analysis*. The algorithm works with clocks to compare time stamps when a shared location is accessed. There, each thread holds a clock and keeps track of all other thread clocks. Moreover, each shared location stores the time stamps of the last accessing thread. If a thread gains access to a shared location, its time stamp must be higher or equal to the time stamp of previously called threads. Otherwise a race condition has occurred. Modern tools, such as MultiRace [PS03] and RaceTrack [YRC05], combine both approaches. The need for an instrumentation prevents an application of these tools in production systems. Furthermore, dynamic race detectors are not sound, i.e. they only check code that is actually executed which is similar to the delta debugging approach. Finally, all tools utilize implementations to handle the very specific error class of race conditions. On the other hand, delta debugging is a more general approach.

The verification of particular properties of concurrent systems is often done by using formal verification techniques, e.g. to show the absence of deadlocks or data races, e.g. [HJM04], [Sto00], or fairness properties. For this reason, a suitable high-level abstraction of the system has to be available.

Either the abstraction is created from the actual system or it is defined directly from the specification which could be a non-trivial and error-prone task. Such an abstraction suffers from the state explosion problem. Hence, the technique does not scale very well in case of complex concurrent systems but if the algorithm terminates, the results are precise and sound. Our approach is fully based upon simulation, and thus can handle arbitrary complex, real-world system designs.

Several tools use static analysis techniques to detect race conditions or deadlocks in software programs or system designs, e.g. [EA03], [FF01], [SBR02]. As already mentioned in Chapter 3, static analysis relies on conservative approximations which result in possibly numerous reported false positives. To solve this problem, the most sophisticated tools, such as RacerX [EA03], combine an interprocedural dataflow analysis with heuristics, statistical analyses, and ranking techniques. So, the analysis reports highly precise results on large, real-world programs.

2 AUTOMATIC ISOLATION OF FAILURE CAUSES

At first, this section summarizes the requirements for an automated isolation of actual failure causes. Then, the delta debugging approach is described.

2.1 Requirements

The basic idea of the presented debug procedure is a systematic test of each difference between a failing and a passing test case, and to check whether the failure still exists. If the failure disappears, the cause for that particular failure, also the actual cause, is found. An algorithm implementing the described procedure shall meet the following requirements:

- *Narrowing down strategy.* A strategy is needed that systematically narrows down the difference between a passing and a failing test case.
- *Rating strategy.* An automated test function has to assess a newly created alternate test case whether the failure has disappeared.

Especially the automated test function can be difficult to implement since each class of failure usually requires another test strategy. So, different test data have to be collected and checked such as executed program statements.

2.2 Methodology

A simple and often used debug procedure is simplification. Simplification removes aspects from a failing test case, e.g. lines in the program input, as long as the observed failure disappears. A more efficient approach is delta debugging, abbreviated *dd*, which was proposed by Zeller and Hildebrandt in [ZH02]. It bases on isolation where the passing as well as the failing test case are modified in parallel. Figure 6.2 sketches the general delta debugging algorithm. It calculates the minimal difference between a passing and a failing test case. Whenever a test fails, the failing test case c_F is "reduced". Additionally, whenever a test passes, the passing test case c_P is "increased". Hence, the algorithm iteratively narrows down the minimal difference between a passing and a failing run. As result, the *dd* algorithm returns a test case pair (c'_P, c'_F), where the difference $\Delta = c'_F \setminus c'_P$ is *1-minimal.*

Definition 13. *A difference between a passing test case* c_P *and a failing test case* c_F *is called 1-minimal if this difference is relevant to produce the actual failure. I.e. adding the difference to the passing test case* c_P *would raise the failure [Zel05].*

The efficiency of the *dd* algorithm is closely related to the result of the *test* function. If all test cases return an *unresolved* test outcome, the number of tests in *dd* is quadratic to the difference between the input test cases $|c_F \setminus c_P|$ since the algorithm successively increases its granularity. In cases of a defined result for each iteration, i.e. *pass* or *fail*, *dd* has a logarithmic complexity since it behaves like a binary search.

Despite the automated procedure to isolate an actual failure cause, a human user is often more creative during debugging. That means, he possibly finds the failure cause faster. Nevertheless, an automated process is less error prone and systematically tests the complete search space. In practice, especially the writing of a simple but significant test function is a crucial part of the whole approach. This function has to implement much implicit knowledge to efficiently find the failure cause.

2.3 Approach Rating

Although an automatic debugging approach sounds promising, there are a number of limitations and problems:

- *Define the test function.* One of the most crucial parts of the *dd* algorithm is the definition of a proper test function. This function has to be especially adapted to the particular analysis task to check for the

Require: set of all test cases C, a test function *test* with $test\colon 2^C \rightarrow \{pass, fail, unresolved\}$, a failing test case c_F with $test(c_F) = fail$, a passing test case c_P with $test(c_P) = pass$

Ensure: $(c'_P, c'_F) = dd(c_P, c_F)$ such that $c_P \subseteq c'_P \subseteq c'_F \subseteq c_F$, $test(c'_P) = pass$, $test(c'_F) = fail$, and $\Delta = c'_F \setminus c'_P$ is *1-minimal*

The *Delta Debugging* algorithm is defined as $dd(c_P, c_F) = dd'(c_P, c_F, 2)$

$$= \begin{cases} c'_P, c'_F & if\ |\Delta| = 1 \\ c'_P, c'_P \cup \Delta_i, 2 & if\ \exists i \in \{1, .., n\} \cdot test(c'_P \cup \Delta_i) = fail \\ c'_F \setminus \Delta_i, c'_F, 2 & if\ \exists i \in \{1, .., n\} \cdot test(c'_F \setminus \Delta_i) = pass \\ c'_P \cup \Delta_i, c'_F, max(n-1, 2) & elsif\ \exists i \in \{1, .., n\} \cdot test(c'_P \cup \Delta_i) = pass \\ c'_P, c'_F \setminus \Delta_i, max(n-1, 2) & elsif\ \exists i \in \{1, .., n\} \cdot test(c'_P \cup \Delta_i) = pass \\ c'_P, c'_F, min(2n, |\Delta|) & elsif\ n \leq |\Delta| \\ c'_P, c'_F & otherwise \end{cases}$$

where $\Delta = c'_F \setminus c'_P = \Delta_1 \cup \Delta_2 \cup \ldots \cup \Delta_n$ with all Δ_i pairwise disjoint, and $\forall \Delta_i \cdot |\Delta_i| \approx |\Delta| / n$ holds. The recursion invariant for dd' is $\text{test}(c'_F) = \text{fail} \wedge \text{test}(c'_P) = \text{pass} \wedge n \leq |\Delta|$

Figure 6.2: General delta debugging algorithm according to [ZH02]

certain failure. There, a too complex implementation could exceed the effort of a manual debug procedure.

- *Find passing and failing runs.* A successful application of delta debugging requires two runs, a failing and a passing run. These runs have to be close enough to be minimized to a single difference. If both runs are too distinct, the search space can become too large and the algorithm fails due to complexity reasons.
- *Define a suitable decomposition strategy.* Every application domain needs to define a suitable decomposition strategy. The designer has to find a suitable element to calculate the difference between two simulation runs, e.g. lines of the program input, or thread switching times. The efficient calculation of the difference has an important impact on the algorithm performance.

- *Check the right failure.* During the delta debugging process many different failing test cases could be artificially created. Hence, the observed failure can differ from the searched failing behavior. To reject these alternate failures, the test function has to evaluate the found failure, exactly. Such a situation could be for instance the simulation time the failure occurred, or a stack trace called if the program had crashed.
- *Locate the defect.* Delta debugging reports only the failure-inducing cause for the observed failure. The algorithm does not point directly to the actual defect. Rather, the result suggests a fix to make the failure disappear. To fix the program, debugging is still needed.

Despite these limitations and problems, delta debugging provides an important aid to automate debugging. It helps the designer to isolate a failure cause after a finite time. Hence, this technique could be a valuable approach to locate hard-to-find defects.

3 AUTOMATIC ISOLATION OF FAILURE CAUSES IN SYSTEMC

The ISOC tool integrates the delta debugging algorithm into the SHIELD debugging environment (see Section 3 on page 83). First, ISOC is exemplarily used for the automatic detection of failure-inducing process schedules. The second part describes the application of delta debugging to isolate failure-inducing simulation input in SystemC designs.

3.1 Debugging Process Schedules

Exploiting the range of description capabilities in parallel, multithreaded SDLs such as SystemC complicates debugging of therewith described system models. Especially, the nondeterministic SystemC scheduler can cause many failures that are difficult to debug manually such as deadlocks, or race conditions. A recording of process activations in a process schedule forms one basis to enable an automatic localization of deadlocks or races in SystemC designs. Hence, the failure-inducing difference between two schedules can be narrowed down, automatically. Therefore, ISOC provides a set of features:

- *Deterministic record/replay.* By using the non-intrusive implementation approach of SHIELD (see Section 3.5 on page 93), ISOC extends the SystemC scheduler by a *record* and *replay* facility for simulation

Table 6.1: ISOC debugging commands

Command	*Options*	*Description*
ptrace	record t_{end}	enable schedule recording until simulation time t_{end} is reached
	replay <*file*>	enable replaying the given schedule file <*file*>
	off	turn of replay/record feature
	status	query the current ptrace state
ldl	n/a	start delta debugging if a failing and a passing schedule are available

runs. Hence, process activations in terms of process schedules can be handled. Here, ISOC provides the debugging command *ptrace* (see Table 6.1) which enables schedule recording and deterministic replaying .

- *Isolating failure causes.* The *dd* algorithm automatically narrows down the failure-inducing minimal difference between a passing and a failing process schedule. The resulting schedule produces a failure if and only if a particular method or thread process is activated at a specific point in time during simulation. ISOC starts the *dd* algorithm by calling the *ldl* command (see Table 6.1).
- *Root-cause analysis.* The system design is debugged while replaying the reported failure-inducing process schedule. Here, the system-level debugging features of SHIELD assists the designer in locating the failure-causing defect.

In Figure 6.3 the particular debug process in ISOC is shown. If the user has detected a failure, SHIELD provides a certain debug strategy in terms of the *dd* algorithm.

First, the simulation is run in the record mode to capture a process schedule that is subsequently tested for pass or fail. As soon as a passing and a failing schedule are available, ISOC proposes to narrow down the failure-inducing difference between both schedules using *dd.* After analysis, the minimal difference is reported and the associated process schedule is used to debug the erroneous simulation. During simulation, all debugging features of SHIELD are available to locate the failure-causing defect.

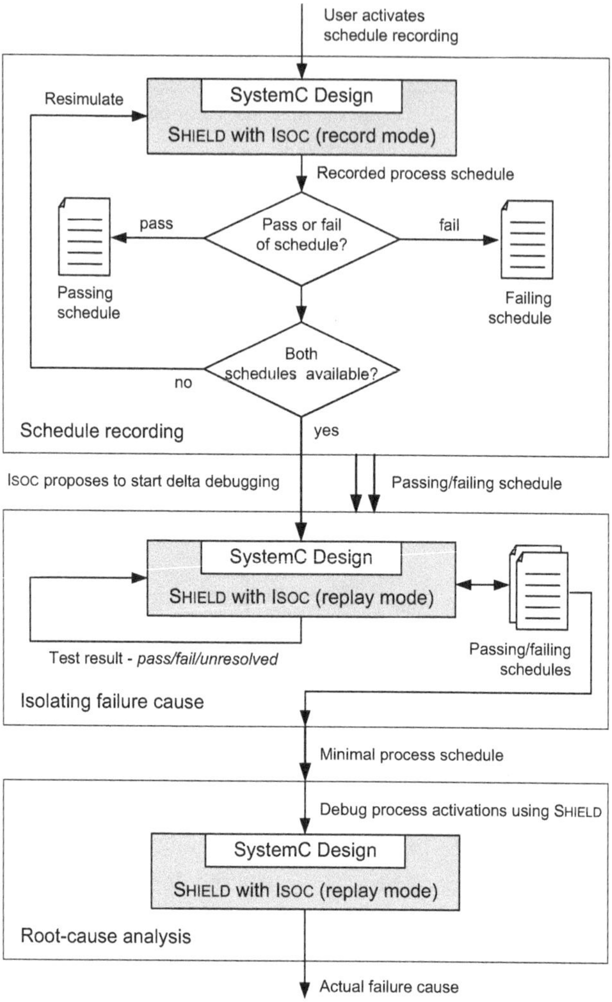

Figure 6.3: Isolation of failure-inducing process schedules

3.1.1 Deterministic Record/Replay Facility

A SystemC design executes all its method and thread processes in a non-preemptive fashion (see Section 1.2 on page 15). Each process activation is determined by the execution logic and is represented by a particular point in time.

Definition 14. *Let D be a SystemC design. A tuple* (t_i, δ_n) *with the simulation time* t_i *with* $0 \leq t_i \leq t_{end}$ *and the* nth *delta cycle* δ_n *with* $n \geq 1$ *is called an activation time point. X is the set of all activation time points of D.*

```
...                              @ 5500 ps
@ 4 ns                           23:
17: C-2                          @ 6 ns
17: D                            25: C-2
17: B-2                          25: A-1
@ 4500 ps                        25: C-1
19:                              25: B-2
@ 5 ns                           25: B-1
21: B-2                          @ 6500 ps
21: C-2                          ...
```

Figure 6.4: Extract of a recorded process schedule

Definition 15. *Let X be the set of all activation time points of a SystemC design D. A 4-tuple* $\Psi = (I, X, \pi, f)$ *is called process schedule of a concrete simulation run of D with*

- I *: the finite set of instantiated method and thread processes in D,*
- X *: the finite set of activation time points of D,*
- π *: a sequence* $\pi = (t_0, \delta_1), \ldots, (t_{end}, \delta_{end})$ *of activation time points, and*
- f*: the function* $f : (t_i, \delta_n) \rightarrow I^*$ *assigning each activation time point a number of processes to be consecutively activated (process activations) at this time point.*

If the record mode is enabled during simulation, ISOC records a process schedule until the specified end time t_{end} is reached.

Example 23. *Figure 6.4 shows an extract of a process schedule recorded for the "deadlock" example presented on page 161.*

A previously recorded process schedule $\Psi = (I, X, \pi, f)$ is replayed during the activated replay mode in five phases:

1. *Initialization phase.* This phase initializes all processes in I and sets the simulation time to the first recorded activation time point in π, i.e. (t_i, δ_n) with $i = 0$ and $n = 1$.
2. *Evaluation phase.* All processes $x \in f(t_i, \delta_n)$ are executed in the recorded order.
3. *Update phase.* This phase performs needed channel updates to propagate data created by previously activated processes.

4. *Delta notification phase.* After delta notifications have been processed, the next element of π is retrieved. In case, processes become active at (t_i, δ_m) with $m > n$, step 2 is called with (t_i, δ_m).
5. *Timed notification phase.* If processes become active at (t_k, δ_m) with $k > i$ and $m > n$, the timed notifications are processed. Moreover, the simulation time is advanced. Then, step 2 is called with (t_k, δ_m). The simulation stops if there is no further element in π.

The *dd* algorithm generates virtual new process schedules without any knowledge of the SystemC simulation and program semantics. So, the consistency of the generated schedule has to be checked for validity. A schedule is discarded and marked with an unresolved simulation outcome, if

- a process is activated twice at (t_i, δ_n) without becoming runnable in that delta cycle a second time,
- the schedule file specifies the execution of a process at (t_i, δ_n) that was already activated at (t_i, δ_{n-m}) although its execution was suspended by a timed waiting statement,
- the simulation logic determines a process that becomes ready to run at (t_i, δ_n) through immediate event notifications but the recorded schedule does not contain a proper activation, or
- a process suspends its execution at (t_i, δ_n) and is waiting for a certain event but the process schedule instructs its activation at (t_{i+k}, δ_{n+m}) without the particular event notification has been occurred.

3.1.2 Isolating Failure Causes

The implementation of the *dd* algorithm as shown in Figure 6.2 is straightforward. Solely, the calculation of the difference Δ between a passing process schedule $c_P = \Psi_P = (I, X_P, \pi_P, f_P)$ and a failing schedule $c_F = \Psi_F = (I, X_F, \pi_F, f_F)$ needs to be defined. To do this, each process activation in f_P and f_F is assigned a unique slot number s_k. These numbers are used to calculate the Δ between the nth activation of a process $x \in I$, written x^n, in both process schedules. Assuming $x^n = s_m$ in c_P and $x^n = s_k$ in c_F. The difference between c_F and c_P for x^n is calculated by $\Delta_P[s_m] = s_k - s_m$, so that $c_P \cup \Delta_P = c_F$. The reduction step $c_F \setminus \Delta_F$ of the *dd* algorithm is implemented by a "reversed delta" Δ_F which is calculated by $\Delta_F[s_k] = s_m - s_k$, so that $c_F \cup \Delta_F = c_P$. To mix both schedules, they have to be possibly aligned using dummy slots. These slots are filled with virtual, never executed, process activations balancing the number of activated processes.

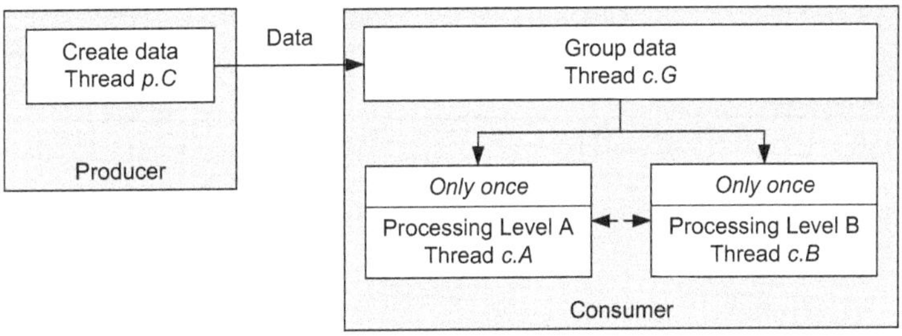

Figure 6.5: Simple producer–consumer application

The *dd* algorithm divides Δ_P and Δ_F into *n* disjoint parts with $\Delta_P = \Delta_{P,1} \cup \Delta_{P,2} \cup ... \cup \Delta_{P,n}$ and $\Delta_F = \Delta_{F,1} \cup \Delta_{F,2} \cup ... \cup \Delta_{F,n}$. Hence, new schedules are iteratively created by $c_P \cup \Delta_{P,i}$ and $c_F \cup \Delta_{F,i}$ where the new target slot of a process activation is calculated by counting back the particular slot difference. The activation time point (t_i, δ_n) of a process is taken from the target schedule, i.e. c_P in case of $c_F \cup \Delta_{F,i}$ and c_F in case of $c_P \cup \Delta_{P,i}$. Due to the partial delta calculation using $\Delta_{P,i}$ or $\Delta_{F,i}$, it could happen that a calculated target slot is already occupied by an unmoved slot entry. In that case, a temporary slot is provided. If the actual slot will be freed due to a further delta calculation, the temporal slot entry becomes the permanent entry. Then, the new schedule is checked for consistency and is simulated to check for pass or fail. Afterwards, the procedure is started again and ends if the difference is 1-minimal.

Example 24. *Figure 6.5 sketches a simple producer–consumer application. The producer component creates a stream of data and sends it to the consumer (SC_THREAD p.C). A thread process at the consumer side (SC_THREAD c.G) receives the data and subsumes them into blocks of a fixed size. After one block is completed, it is sent to a processing unit. Data processing is implemented using two levels, i.e. Processing Level A (SC_THREAD c.A) and Processing Level B (SC_THREAD c.B). A wrong synchronization of the processing threads yields to incompletely processed data in some cases. So, delta debugging is used to locate the defect. Table 6.2 shows the calculation of* Δ_F *(column 4) for the two recorded passing and failing process schedules* c_P *and* c_F *(columns 2 and 3). Both process schedules are represented by a sequence of activation time points and the assigned processes according to Definition 15. Since the failing schedule misses two process activations, two dummy slots are added. Using the first delta part* $\Delta_{F,1}$ *(column 5) generates a new process schedule (column 7) that is simulated (column 8). Here, the dummy slots are left out from any simulation. Due to an erroneous simulation logic the algorithm will proceed with* $\Delta_{F,2}$.

Table 6.2: Illustrating the first steps of the *dd* algorithm

s_k	c_P	c_F	Δ_F	$\Delta_{F,1}$	$\Delta_{F,2}$	$c_F \cup \Delta_{F,1}$	c_{new}
0	$((0,1), c.G)$	$((0,1), c.G)$	0			$((0,1), c.G)$	$((0,1), c.G)$
1	$((0,1), p.C)$	$((0,1), p.C)$	0			$((0,1), p.C)$	$((0,1), p.C)$
2	$((0,1), c.B)$	$((0,1), c.A)$	1	1		$((0,1), c.B)$	$((0,1), c.B)$
3	$((0,1), c.A)$	$((0,1), c.B)$	−1	−1		$((0,1), c.A)$	$((0,1), c.A)$
4	$((0,1), p.C)$	$((0,1), p.C)$	0			$((0,1), p.C)$	$((0,1), p.C)$
5	$((0,1), c.G)$	$((0,1), c.G)$	0			$((0,1), c.G)$	$((0,1), c.G)$
6	$((0,1), p.C)$	$((0,1), p.C)$	0			$((0,1), p.C)$	$((0,1), p.C)$
7	$((0,1), c.G)$	$((0,1), c.G)$	0			$((0,1), c.G)$	$((0,1), c.G)$
8	$((0,1), c.A)$	$((0,1), c.B)$	1	1		$((0,1), c.A)$	$((0,1), c.A)$
9	$((0,1), c.B)$	$((0,1), c.A)$	−1		−1	$((0,1), c.B)$	$((0,1), c.B)$
10	$((0,1), c.A)$	$((20,3), c.B)$	1		1	$((20,3), c.B)$	$((20,3), c.B)$
11	$((20,3), c.B)$	$[((20,4), c.A)]$	−1		−1	$[((20,4), c.A)]$	
12	$((20,3), c.A)$	$[((20,4), c.A)]$	0			$[((20,4), c.A)]$	

Besides the delta calculation, a test function *test* has to be implemented. This function corresponds to the particular analysis task, e.g. deadlock detection or data race isolation. ISOC supplies a test function template that has to be implemented by the designer according to the design analysis problem.

3.1.3 Root-Cause Analysis

Finally, the reported minimized process schedule is replayed using ISOC. The system-level debugging features of SHIELD (see Section 3.3 on page 86) allow to debug the location of the defect that has caused the actual failure. There, simulation time and delta cycle information in the schedule file provides valuable information for defect localization.

3.2 Debugging Program Input

Delta debugging can be also used to automatically isolate a certain piece of SystemC simulation input yielding an erroneous simulation behavior. The objective is to find the minimal difference between two input data sets. The minimal difference could be a single character (set), or a particular line of input. As basis for the realization an existing Python implementation of the *dd*

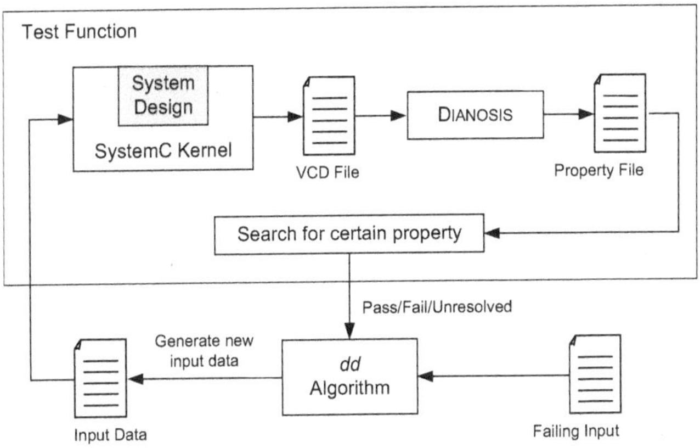

Figure 6.6: Isolating failure-inducing simulation input

algorithm is used [Zel08]. This implementation was modified to determine the difference between two text files on a line-by-line basis instead of a decomposition of the input into single characters. The test function is implemented by analyzing properties generated by DIANOSIS (see Section 3 on page 116). Hence, missing or additional properties make the difference between a passing or a failing test. Figure 6.6 depicts the described analysis flow.

4 EXPERIMENTAL RESULTS

The following section summarizes the experimental results while applying the *dd* algorithm to different SystemC designs. First, delta debugging is used to automatically detect a failure-inducing instruction sequence in erroneous SIMD programs. Second, two experiments demonstrate the usage of ISOC to narrow down failure-inducing process schedules.

4.1 SIMD Data Transfer Example Continued

In Chapter 5 1,009 randomly created programs were simulated on the SIMD data transfer example to produce 1,009 simulation traces. Thirteen of these programs have shown a difference in the number of generated properties. Precisely, the difference concerned the handshake property `core.ld_req |-> core.st_req`. As already shown in Chapter 5, this finding has indicated an erroneous handling of the instruction sequence "*multiply–subtraction instruction* followed by a *maximum calculation*". There, the error was manually detected by a comparison of the commonalities between all 13 programs. This manual procedure is a very tedious and error-prone.

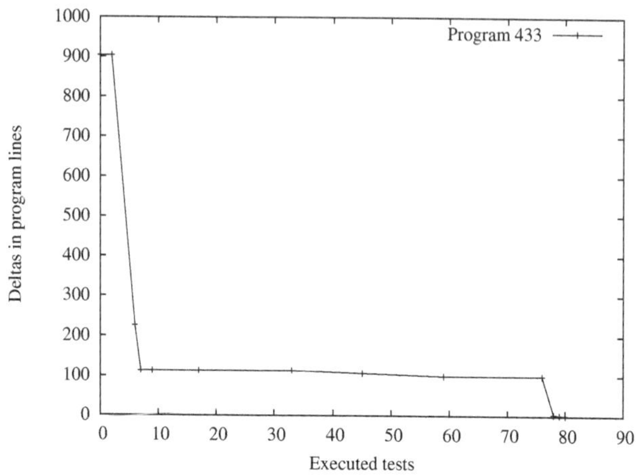

Figure 6.7: Isolate the failure-inducing instruction in program 433

By using delta debugging, the failure-inducing instruction sequence can be isolated automatically. Here, the analysis flow from Figure 6.6 is applied. The *test* function checks whether the missing handshake property could be found or not. The *dd* algorithm is started with the failing program while the empty program is assumed to pass. As result, the algorithm reports the minimal difference between a passing and a failing SIMD program in terms of a single instruction. So, either the *multiply–subtraction* or the *maximum calculation* instruction has to be removed to make the failure disappear, e.g.:

```
Process program tmp/prog_116
Output: Error detected: [(191, 'maxu R15, R15, R1\n')]
Done. See tmp/prog_116.delta.log1 and tmp/prog_116.delta.log2
for details

Process program tmp/prog_204
Output: Error detected: [(21, 'mulsubss R7, R0, R11, R3\n')]
Done. See tmp/prog_204.delta.log1 and tmp/prog_204.delta.log2
for details
```

Using an automated approach allows a much faster error location. The interpretation of the analysis output of all 13 programs suggested that either one of the reported instructions or their combination result in the observed wrong simulation behavior.

Figure 6.7 depicts the *dd* algorithm while narrowing down the failure-inducing program statement in program 433. After 80 tests, 904 initial differences are reduced to a minimal difference.

Table 6.3: *dd* algorithm applied on 13 erroneous programs

Program	*Initial difference (# program lines)*	*Overall test runs*	*Passed tests*	*Failed tests*	*Unresolved tests*	*Analysis time*[a]
116	329	151	4	6	141	12.5 s
204	406	340	3	21	316	32.4 s
208	400	479	3	47	429	47.0 s
433	904	80	4	3	73	9.3 s
440	391	27	5	2	20	3.9 s
548	623	312	4	9	299	28.0 s
669	726	984	17	103	864	1m 51.8 s
755	329	202	3	13	186	17.2 s
787	281	414	3	37	374	38.9 s
788	752	1,198	19	138	1,041	2 m 22.2 s
856	705	1,076	10	97	969	2 m 3.0 s
861	403	398	3	37	358	36.8 s
981	938	258	3	6	249	24.0 s

a. *Test system:* AMD Opteron™ 248 processor @2200 MHz, 3 GB RAM

Table 6.3 summarizes the number of required test runs to isolate the failure-inducing instruction sequence in each of the 13 erroneous programs. The *dd* algorithm requires 27 to 1,198 runs while this number does not correlate with the length of the input program, i.e. the initial difference. This discrepancy results from the very different number of unresolved tests while checking new programs. New programs are created by mixing the current failing and passing program. There, a different quantity of spurious programs is generated that violate the program logic. The violated program logic produces an unresolved test outcome. The analysis times, that are reported in Table 6.3, show the efficiency of an automatic approach. In such a short time a manual debug procedure would not be possible, in general. After the failure-causing input is known, the example is debugged using the SHIELD debugging environment. As a result, the designer identifies a wrong synchronization of the data transfer in case of a combination of both instructions.

```
### PASSING RUN ###                 ### FAILING RUN ###

@ 0 ns                              0 ns
1: p.C                              1: p.C
1: c.G                              1: c.G
...                                 ...
>>>>>>>>>>>>>>>>>>>>>>>>>>>>>>>>>>>>>>>>>>>>>>>>>>>>>>>>>>>>>>>>>>>>>>>>>>>>>>>>
1: c.A                              1: c.B
1: c.B                              1: c.A
1: c.A
@ 20 ns                             @ 20 ns
3: c.B                              3: c.B
3: c.A
```

Figure 6.8: *dd* algorithm analysis result for Example 24

4.2 Failure Causes in Process Schedules

This section summarizes the experimental results concerning the isolation of failure-inducing process schedules (see Section 3.1 on page 150). Two synthetic SystemC designs were analyzed, i.e. a simple producer–consumer design and a deadlock example. Both examples randomly, but definitely, produce various erroneous situations.

4.2.1 Producer–Consumer Application

In general, a SystemC design has to be independent of any process execution order. In the producer–consumer application from Example 24, a wrong synchronization could lead to wrong simulation results. Hence, transmitted data are incompletely processed. Here, the nondeterminism at the delta cycle level causes a serious problem: If the data processing thread `c.A` is initially executed before thread `c.B`, the initial event notification runs into the void since events are not persistent in SystemC. The wrong simulation outcome is used to write a proper test function for the *dd* algorithm. Following the flow in Figure 6.3, a passing and a failing run are generated at first. After 21 test runs, the *dd* algorithm has been isolated the failure-inducing minimal difference between both recorded process schedules. Figure 6.8 depicts the *dd* analysis result. The marker line indicates the (first) difference between both process schedules. Hence, the user knows that the failure is caused by a changed activation sequence of the threads `c.A` and `c.B`. Using the reported failing schedule allows to replay and debug the erroneous simulation in ISOC. Debugging shows that the event used for thread communication is sent by thread `c.A` but runs into the void since there is no process waiting for it, yet.

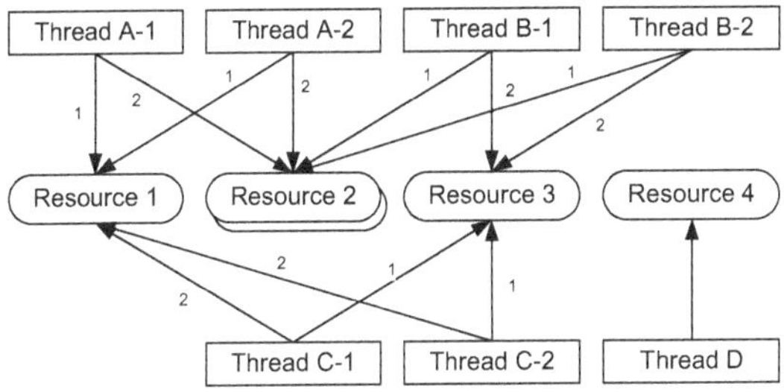

Figure 6.9: Architecture of the deadlock example

4.2.2 Deadlock Example

Figure 6.9 sketches the general architecture of a SystemC design definitely producing a deadlock after a random time. The example is based on a SystemC Training Course [IIS]. Four types of threads, i.e. A, B, C, and D, try to acquire four different resource types, i.e. 1 to 4. The design provides two instances of resource 2 and one instance of all other resources. During simulation each thread instance tries to lock two particular resource instances one after another to start "working". If the thread gets the first resource, it tries to lock a second resource, immediately. In case a lock does not succeed, the thread waits for a random time and tries a relock. If both resources could be successfully locked, the thread "works" a random time and releases the resources afterwards. Since the thread keeps the first resource locked, while it is waiting for the availability of the second one, a deadlock will randomly occur.

To isolate the process which causes the observed deadlock, a test function for the *dd* algorithm is needed. Here, a resource allocation graph is created from a generated simulation log. Using the graph, a deadlock can be detected, whenever a cycle is found. Since the example can produce various deadlock situations at random, the particular delta cycle and the simulation time of the actual deadlock is saved. Hence, the test function looks only for that particular deadlock situation (see Section 2.3 on page 148). However, a simple test function would recognize any deadlock whenever two threads are waiting for each other.

For test purposes, the example was run several times to create passing and failing process schedules over different simulation time lengths. Table 6.4 summarizes the obtained analysis results running the *dd* algorithm on the example with a clock period of 1 ns.

The second column shows the initial difference between passing and failing process schedules. After constructing a virtual new schedule, it is written into a file and is simulated, afterwards. Either the new schedule results in a

Table 6.4: Running *dd* on the deadlock example

Simulation time in ns	*Initial difference between process activations*	*Overall test runs*	*Passed tests*	*Failed tests*	*Unresolved tests*	*Analysis time*[a]
50	2,820	48	2	17	29	7 s
100	16,330	158	6	17	135	29 s
200	51,594	335	6	22	307	68 s
300	51,972	218	4	33	181	49 s
400	147,290	825	4	75	746	193 s
500	155,322	858	4	61	793	230 s
1,000	1,071,436	363	4	70	289	204 s
2,000	7,716,550	617	11	202	404	1,803 s

a. *Test system:* Intel Centrino Duo T2400@1830MHz, 1 GB RAM, SystemC Kernel 2.2.0 compiled with gcc-4.2.4

passing (column 4) or a failing test outcome (column 5). Due to a violated simulation semantics, unresolved test outcomes (column 6) are produced most frequently. As can be already seen in Table 6.3, the number of needed simulation runs does not correlate to the initial difference between the input schedules. In fact, the number of unresolved test outcomes has a major impact on the performance of the *dd* algorithm (column 7). In the current implementation the *dd* algorithm and the SystemC simulation communicates via files which hampers an efficient error search for long running design simulations. Nevertheless, Table 6.4 demonstrates the applicability of the *dd* algorithm to isolate failure-inducing process activations in SystemC designs systematically in a finite and short time.

5 SUMMARY AND FUTURE WORK

In this chapter an experimentation technique has been proposed that automatically isolates various failure causes in ESL designs written in SystemC. The delta debugging algorithm narrows down failure causes using a series of controlled simulation runs. The algorithm reduces a passing and a failing run to a minimal difference automatically. There, all parts irrelevant for that failure are removed. The reported difference causes the observed failure and suggests a fix to make the failure disappear.

Two use cases of delta debugging demonstrate its applicability for ESL design and emphasize the strength of a combination of different debugging techniques. The first use case shows the isolation of a failure-inducing simulation input. Therefore, property generation as inductive technique has been used to write a proper test function. The second use case extends an observation technique in terms of the SHIELD debugging environment by the ISOC component. ISOC provides a flow to find the minimal difference between two process schedules.

Depending on the application, failures caused by a different execution order of processes can be automatically found. As a basis, the SystemC scheduler was augmented by a deterministic record and replay facility. Herewith, formerly recorded simulation runs can be replayed. Based on two input schedules, ISOC generates alternate process schedules until a minimal difference between the input schedules has been found. Subsequently, the debugging features in SHIELD and the replay facility of ISOC support the designer to locate the actual defect.

Summarized, the presented approach supports an automatic debugging of complex system designs especially in cases where the designer has no clue how to start debugging. An experienced designer sometimes finds the failure cause in a fewer steps. In contrast, the automated approach, although dump, completely and systematically tests the search space and will come up with a result in any case. Three experiments have been documented the efficiency of the *dd* algorithm. In all examples the failure-causing minimal difference could be narrowed down after a relatively short time. So, the designer gets a result in any case which is (usually) reported in less time than a manual procedure would take.

Chapter 7

Summary and Conclusion

ESL design promises to overcome the limitations of designing complex integrated circuits and systems at the RTL. The related system model allows for an early system exploration and verification. This model is used as a golden reference for subsequent design stages. So, its correct and reliable creation is a key issue for successful ESL design. Currently, ESL verification consists of a large variety of techniques such as simulation, semi-formal verification, and formal verification. Although formal approaches get an increasing attention and application, simulation is still the predominant verification technique to ensure the functional design correctness. Design concepts like object-orientation, borrowed from software development, and TLM have boosted the design productivity at the ESL over the last years. In parallel, the verification productivity falls behind whereas pinpointing the failure-causing bug still remains an unsystematic and time-consuming process. So, it becomes important not only to find many errors early during development. Rather, new methods and approaches have to be provided that improve and support an automation of debugging.

In this book, a debugging approach for ESL designs has been proposed. This approach accompanies the various development stages of the system model systematically. Particular debugging techniques form a hierarchy of co-ordinated reasoning techniques that minimizes the number of errors escaping to the next development stage. All techniques were empirically evaluated in case studies and experiments mostly using real-world industrial designs. There, a special focus is taken onto SystemC. Additionally, an example, that is used throughout the book, demonstrates the strengths and particularities of each technique. Here, especially the integration aspect is highlighted.

As soon as first syntactic correct modules of the system model are available, static analysis enables an early detection of errors. The code quality is ensured by checking coding standards statically. Corporate coding standards are a major issue in distributed development teams and for a higher level IP reuse process. Moreover, critical coding flaws and typical pitfalls can be identified and fixed. For static analysis a framework and the framework-based SystemC analyzer SDAS has been introduced. Using SDAS, the designer can ensure a high code quality from the very first implemented lines of code. So, the code contains fewer bugs before a first subsystems become simulatable. If a design is simulated the first time, there is a great chance to find many new

F. Rogin and R. Drechsler, *Debugging at the Electronic System Level*,
DOI 10.1007/978-90-481-9255-7_7, © Springer Science+Business Media B.V. 2010

bugs. A debug flow has been introduced that allows a debugging and exploration of system models at a higher abstraction level. Based on that flow, the SHIELD debugging environment for SystemC was presented. The SHIELD environment eases the detection, identification, and location of errors directly at system level. The aggregation of multiple simulation runs offers new opportunities to learn various aspects about the design. A new methodology for an automatic generation of complex design properties has been proposed and prototypically implemented by the DIANOSIS tool. Inferred properties help to indicate anomalies that point to gaps in the test suite or errors in the design description or the specification. However, more important is the contribution of property generation to an improved design understanding through the interactive designer's review and discussion of found properties. Hence, a new and promising approach to check the implemented design against the textual specification is supplied. Determining the actual failure cause is still a manual, error-prone, and tedious procedure. So, a technique that automatically narrows down the failure-inducing minimal difference between a passing and a failing test case in SystemC designs was investigated. The tool ISOC implements the delta debugging algorithm. This technique extends the debugging features of the SHIELD environment by an automated approach that supports the designer in detecting hard to find errors.

Comparing the efficiency of a design flow using the proposed debugging techniques with a usual design flow without these techniques is difficult. In summary, the techniques systemize and accelerate error detection, observation, and isolation as well as design understanding. So, the time between the insertion of an error and its correction is minimized. Illustrated at the SIMD data transfer example, that is used throughout the book, this statement is emphasized by the following facts:

- Using static analysis, a functional error in a SIMD design component was found and fixed long before the system model becomes simulatable (see Section 4.1 on page 63).
- While the SIMD example is simulated the first time, a failure was observed. This failure was quickly traced back to a copy-paste error. Here, a debug pattern in connection with only three system-level debugging commands and a proper visualization accelerated error detection (see Section 4.1 on page 95).
- Property generation, applied over multiple simulation runs, has inferred an important data transfer property in the SIMD design already at system level. This property must remain valid during all following design stages up to the final hardware implementation.

Thus, it is formulated as an assertion and checked during further development (see Section 4.1 on page 129).

- A failure-inducing instruction sequence has been automatically isolated in 13 erroneous SIMD programs within an analysis time of 3.9 s to 142.2 s performing controlled simulation runs. The achieved runtimes are much smaller than a manual debug procedure would take. The reported failure cause points to an erroneous handling of a rare instruction sequence (see Section 4.1 on page 157).

The efficiency of static analysis for error detection has been already shown by several empirical studies. So, 60% of software defects found in released products could have been detected by means of static analysis [PC95]. Static analysis techniques usually scale very well to large code bases of million lines of code, e.g. [EC+01], or [Das00]. Since a system model description is also pure code, a static analysis of such descriptions should come to similar results. The SystemC analyzer SDAS checks about 300,000 lines of code in several seconds. The found coding violations simplify manual code reviews and help the designer to follow a corporate coding standard.

The SHIELD debugging environment uses the GDB debugger which is utilized successfully in many development environments such as *Eclipse* [CDT], *Real View Debugger Suite* [ARM], or *Platform Architect* [CoWare]. Since a debugger usually observes a small piece of code at a particular point in time, it shows a good scalability in terms of large code bases. Moreover, SHIELD provides system-level debugging commands as well as visualization features. These features reduce the number of required commands significantly to reach a point of failure, or a particular simulation state. Lastly, many concepts of SHIELD have been developed and proven in close co-operation with a large industrial partner ensuring their industrial applicability.

The generation of complex properties using DIANOSIS provides a whole new verification methodology. This methodology facilitates property creation and improves design understanding. The new approach has shown its effectiveness on real-world industrial designs.

Finally, the usage of the delta debugging algorithm is an important and new contribution to automate SystemC debugging, especially in case of hard to find errors. The minimal difference between a passing and a failing test points to a failure cause that is normally found after a short and finite time. Moreover, the reported cause is used to find the failure-causing defect, more quickly.

Summarized, all introduced debugging techniques have shown their efficiency and scalability in several experiments. The techniques accompany the overall verification process of a system model from first modules to the final model. So, the proposed debugging approach makes important contributions to a systematic and automatic debugging of ESL designs. Finally, it results in an improvement of the verification productivity at system level.

Appendix A

FDC Language

The following chapter defines the syntax and the (informal) semantics of the FDC language. FDC specifications are the common configuration and implementation approach in the REGATTA analysis framework. They are used to adapt the generic framework components to a concrete language. Moreover, these specifications allow to define various analyses, e.g. implementing arbitrary coding checks.

1 FDC SYNTAX

The notation of FDC specifications is geared to the syntax of the *State Machine Compiler* [Rapp08]. In Figure A.1 an EBNF of the FDC language is given. This allows to describe abstract and configurable analysis tasks.

2 FDC SEMANTIC

The informal semantic is specified as follows:

- A FDC specification divides into seven sections:
  - * The *require* section contains a list of C++ class header files to be included.
  - * The *import* section contains a list of imported variables.
  - * The *variable* section consists of any number of locally used variables.

```
fdc_spec        ::= {fdc_desc}+
fdc_desc        ::= t_FDC ident '{' {require_decl}* {import_decl}* {var_decl}*
                                    [prior_decl] [init_decl] {func_decl}*
                                    {state}+ '}'
require_decl    ::= t_REQUIRE <include_path>
import_decl     ::= t_IMPORT type ident ';'
var_decl        ::= t_VAR var_type ident ';'
prior_decl      ::= t_PRIORITY number ';'
init_decl       ::= t_INIT '{' action '}'
func_decl       ::= t_FUNCTION ident '{' action '}'
state           ::= state_name [entry] [exit] '{' {trans}+ '}'
trans           ::= proxy [ '[' guard ']' ] state_name [ '{' action '}' ]
proxy           ::= ident
state_name      ::= ident | t_NIL
entry           ::= t_ENTRY '{' action '}'
exit            ::= t_EXIT '{' action '}'
type            ::= t_INT | t_STRING | t_DOUBLE | t_BOOL
ident           ::= [a-zA-Z_][a-zA-Z0-9_]+
var_type        ::= // any correct C++ type
action          ::= // correct C++ code + Regatta API
guard           ::= // correct C++ Boolean expression
```

Figure A.1: EBNF syntax of the FDC language

  * The *priority* section specifies the particular priority of the FDC specification.
  * The *init* function represents the FDC initialization routine.
  * The *function* section defines C-like functions which can be called from any action code.
  * The *FSM* section describes the underlying FSM used to control the analysis steps.

- Each `require` statement specifies the name of a header file to be included. It is used if the action code references functionality that is not included per default by the C++ compiler. Include files are searched in all REGATTA source code paths and the standard include paths.
- The `import` statement declares a variable that has to be externally set in the FDC XML configuration file with respect to a concrete analysis task.
  NOTE: Imported variables are read only in all action code blocks otherwise the compiler reports an error.
- A `var` statement declares a local variable that can be read or written in any action code block of the FDC specification.
- Allowed types in an `import` and `var` declaration must satisfy C++ syntax for type declarations.

- The priority of an FDC can be specified by the `priority` statement. This statements defines a call sequence if two FDCs have registered for the same elements of the created EST (Definition 4). As higher the given priority value is as later the FDC is called.
- The `init` function is executed just before syntax analysis starts. It may contain arbitrary initialization code for the FDC using correct C++ syntax.
- A `function` declaration defines a function with no return value and an empty argument list. It may contain any correct C++ code and is called within the transition action code segments like a usual C function.
- The actual FSM specification consists of any number of states and an arbitrary number of transition between these states.
  * A *state* specifies any number of transitions that switch the FSM into a new state. The special state name `nil` is reserved for a transition leading back to the current state. A state may have one `entry` and one `exit` function. Code in the `entry` function is executed when the state is newly entered. The `exit` code block is evaluated when the current state is left. The lexically first state implicitly represents the initial start state.
  * A *transition* consists of a *proxy symbol*, a *guard*, and a *target* state. A proxy symbol is a wildcard for elements of the analysis alphabet of the given context-free grammar. Elements of a proxy symbol trigger the particular transition. The target state has to be the name of another existing state inside the FDC specification. A conditional transition can be implemented by using an optional guard. The guard is evaluated during FDC execution and only if it evaluates to *true* the transition is triggered. A guard must be a valid C++ Boolean expression and may contain queries of the current state of another FDC specification. Such a query has to meet the following syntax:

    ```
    // <FDC name>. name of the target FDC
    // <queried state>. existing state in the target FDC
    <FDC name>->getState() == <queried state>
    ```

  * An action code block has to contain any correct C++ code. Per default, the user has the full access to the REGATTA API using two implicit variables: `AM_AnalysisMngr* am` which allows to access any framework components, and `FE_Data* data` that refers to the current language-specific EST element. The fol-

lowing example sketches the access to a meaningful token (see Section 1.1 on page 34):

```
function check
{
   // retrieve the next meaningful token
   FE_Token* tok = data->getNextLRT();
   if (!tok) {
      return;
   }
   std::count << "Token name: "
               << tok->getName() << std::endl;
}
```

Appendix B

Debug Pattern Catalog

At first, this chapter describes the general description format used for the debug pattern catalog. Subsequently, two debug patterns are introduced, exemplarily.

1 GENERAL FORMAT

Each debug pattern is described in the following common format. Such a format is already used by Gamma et al. [GHJV95] to describe their object-oriented design pattern catalog:

- *Pattern name.* The format description is introduced by a short and intuitive name for the debug pattern.
- *Motivation.* The existence of the debug pattern is motivated in this paragraph. Here, a representative problem is illustrated.
- *Symptom.* This paragraph describes the failure symptom which is typical for the error the pattern is aimed at.
- *Assumption.* The assumption specifies the problem or certain conditions, i.e. a specific architecture or component composition in the design, that probably causes the observed symptom.
- *Participants.* All participating system-level debugging commands and their specific responsibilities in the pattern context are described in this paragraph.
- *Debug procedure.* The debug procedure is described by a flowchart to formally document the necessary steps to isolate the error.

- *Example.* This paragraph sketches a typical situation where the pattern helps the designer to find and correct an error.
- *Related patterns.* Related patterns also match the observed failure symptom. These patterns should be tried if the failure-causing defect could not be found using the current pattern.

2 COMPETITION PATTERN

Motivation

A typical SystemC design consists of different communicating processes that are running in parallel and that are synchronized over simulation time. The execution order of processes is not determined during a simulation delta cycle. This introduces nondeterminism into the simulation process and may lead to unpredictable and wrong design behavior. Examples for such wrong design behavior are race conditions or deadlocks.

Symptom

The design is simulated (with the same inputs) and unexpectedly produces different output or erroneous results, possibly in some minor cases. This situation could be caused by a nondeterministic execution of parallel processes in the same delta cycle or by an erroneous process synchronization.

Assumption

There are at least two processes concurrently competing for the same (shared) resource such as a signal or a bus.

Participants

dp_sense	return all processes triggered by the same event
lpt_rx	review the sensitivity list of the given process
lst	explore the source code of a process
lse_rx	get the correct hierarchical event name

Debug Procedure

Figure B.1 shows the debug procedure of the COMPETITION pattern.

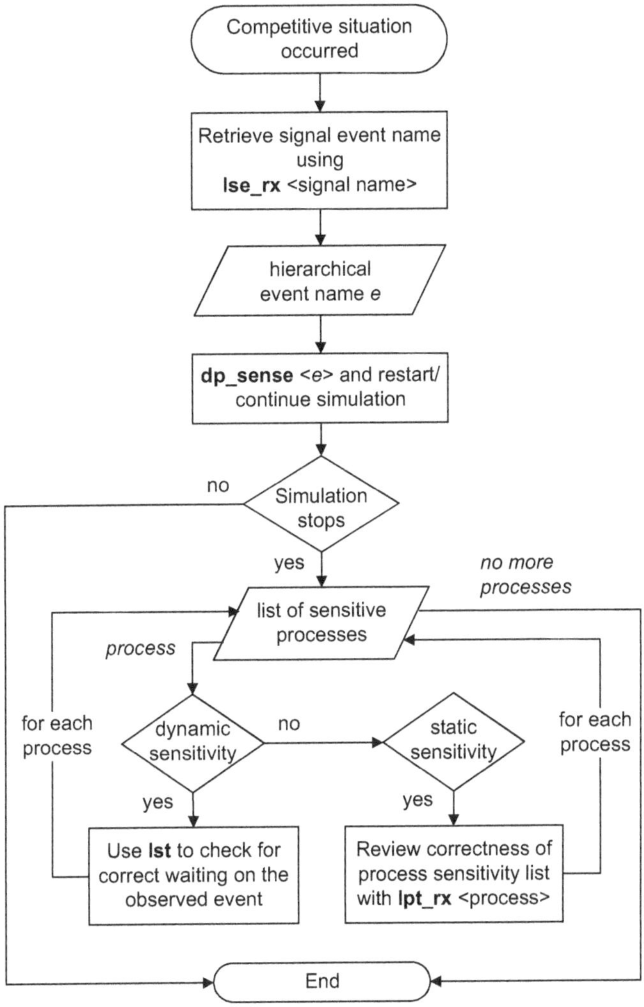

Figure B.1: COMPETITION pattern flowchart

Example

Problem. Figure B.2 sketches a situation where two thread processes *top.bfm._fsm_update* and *top.bfm._fsm_rst* write to the same signal *top.ctrl_w*. Both threads are activated by the ready signal *top.rdy_l*. Since the SystemC simulation kernel does not define a deterministic order of thread activations inside a delta cycle, a race condition can occur on signal *top.ctrl_w*.

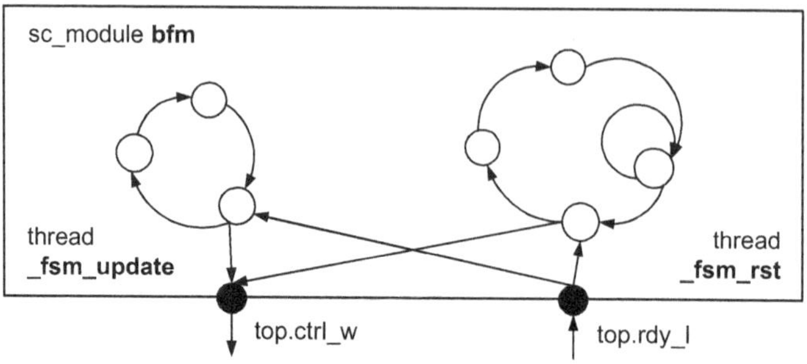

Figure B.2: Exemplary race condition

Debug procedure. Assume that the hierarchical event name of the signal causing the problem is already known. Thus, *dp_sense* is called with it and the simulation is restarted:

```
(gdb) dp_sense "top.rdy_l.m_negedge_event"
*** COMPETITION debug pattern activated
*** restart/continue simulation using run/continue
(gdb) run
```

When the simulation stops *dp_sense* reports two thread processes sensitive to the observed event. For convenience reasons, the debugging environment additionally adds breakpoints at the sensitive processes:

```
*** dp_sense 'top.rdy_l.m_negedge_event'
*** Check competitive situation between sensitive processes
  in module top.bfm
     top.bfm._fsm_update  <static>
     top.bfm._fsm_rst  <static>

*** breakpoints in sensitive processes
  breakpoint at top.bfm._fsm_update
  breakpoint at top.bfm._fsm_rst
```

Knowing that thread *_fsm_update* is correctly activated by the ready signal, the sensitivity list of the statically triggered thread *_fsm_rst* is investigated using *lpt_rx*:

```
(gdb) lpt_rx "top.bfm._fsm_rst"
process top.bfm._fsm_rst sensitive to
   <static> top.fsm_rst_l.m_negedge_event
   <static> top.rdy_l.m_negedge_event
   <dynamic> top.write_tx.m_value_changed
```

The command shows that the sensitivity list falsely includes the ready signal which turns out to be an environment defect.

Related Patterns

LIVELOCK

3 TIMELOCK PATTERN

Motivation

Event-based communication between concurrent processes can often lead to a lock condition. Then, a process is caught in an infinite loop and thus never waits for an event.

Symptom

The simulation infinitely loops or at least appears to do so. Additionally, the simulation time does not proceed.

Assumption

Due to design specification knowledge, the user suspects one or more processes causing the lock.

Participants

dp_timelock	manually look for hanging processes
lst	explore the source code of a process
dstep	proceed simulation step-wise

Debug Procedure

Figure B.3 shows the debug procedure of the TIMELOCK pattern.

Example

Problem. Figure B.4 illustrates a system where a device (*top.i_device*) communicates with a host (*top.i_host*) over a bus. Device and host exchange data using the two signals *top.req_data* and *top.resp_data*. Available data packages are indicated by two ready signals *top.req_ready* and *top.resp_ready*, respectively. The user models the signal interaction in a faulty way. So, the simulation locks.

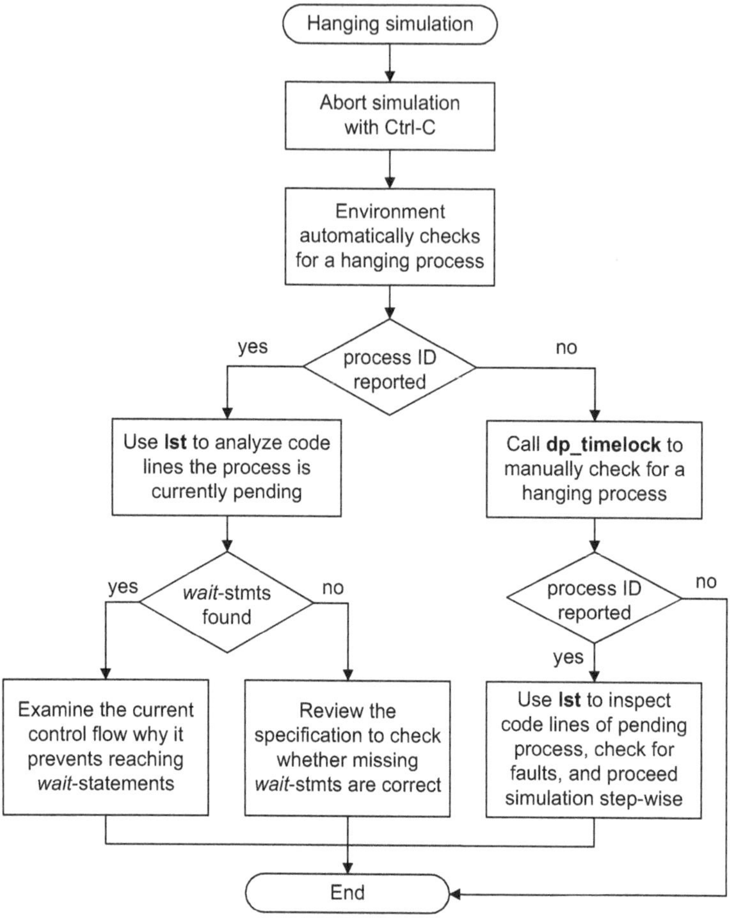

Figure B.3: TIMELOCK pattern flowchart

Debug procedure. The user aborts the simulation by pressing Ctrl-C. It is assumed, that the debugger does not report any process ID. Hence, the *dp_timelock* command is called to manually check for a hanging process:

```
Program received signal SIGINT, Interrupt.
(gdb) dp_timelock
*** following process seems to hang
top.i_device._rx_tx
*** TIMELOCK debug pattern activated
*** call lst 'top.i_device._rx_tx' to examine source code of
pending process
```

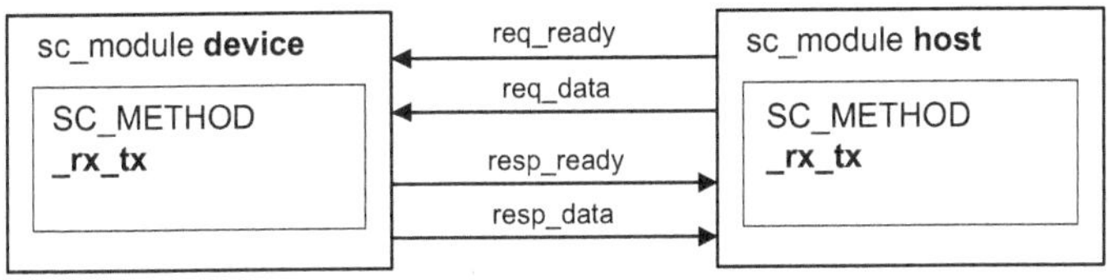

Figure B.4: Exemplary TIMELOCK pattern

The source code of the hanging process is investigated to find the root cause of the observed failure:

```
(gdb) lst 'top.i_device._rx_tx'
--lst: list active source of [c]thread/method---
process top.i_device._rx_tx is currently
 at /home/hld/project/tb/src/device.cpp:254
in device::_rx_tx
254          void device::_rx_tx() {
255              process_req_data(req_data.read());
256              resp_data.write(gen_resp_data());
257              resp_ready.write(~resp_ready.read());
258
```

At first glance, no error is found. Thus, the simulation is proceeded in step-mode using multiple *dstep* commands:

```
(gdb) dstep
*** setting breakpoint(s) at next delta cycle
*** runnable process(es) at delta cycle 3563, @14 ns
   break @ 'host::_rx_tx()' of 'top.i_host._rx_tx'
*** type <continue> to visit breakpoints
(gdb) cont
```

After some more *dstep* commands, the debugger log shows that the processes *top.i_host._rx_tx* and *top.i_device._rx_tx* become mutually active, invoking each other without any simulation time progress. A code review of the affected modules identifies a lack of time-consuming statements within the thread loop.

Related Patterns

DEADLOCK, LIVELOCK

References

[Actis] Actis Design. AccurateC: Static C++ Code Analysis for SystemC. *Actis Design*, Portland, 2008.

[AB+07] A. Aiken, S. Bugrara, I. Dillig, T. Dillig, B. Hackett, and P. Hawkins. An Overview of the Saturn Project. In *Workshop on Program Analysis for Software Tools and Engineering*, pp. 43–48, 2007.

[ABL02] G. Ammons, R. Bodik, and J.R. Larus. Mining Specifications. *ACM SIGPLAN Notices*, Volume 37, Issue 1, pp. 4–16, 2002.

[ABS03] G. Agosta, F. Bruschi, and D. Sciuto. Static analysis of transaction-level models. In *Conference on Design Automation*, pp. 448–453, 2003.

[AF02] T. Arts and L.-A. Fredlund. Trace Analysis of Erlang Programs. In *ACM SIGPLAN Workshop on Erlang*, pp. 16–23, 2002.

[AP+07] N. Ayewah, W. Pugh, J.D. Morgenthaler, J. Penix, and Y. Zhou. Evaluating Static Analysis Defect Warnings on Production Software. In *Workshop on Program Analysis for Software Tools and Engineering*, pp. 1–8, 2007.

[ARL00] D. Abts, M. Roberts, and D.J. Lilja. A Balanced Approach to High-level Verification: Performance Trade-offs in Verifying Large-scale Multiprocessors. In *International Conference on Parallel Processing*, pp. 505–510, 2000.

[ARM] ARM Ltd. RealView Development Suite Homepage. [Online], http://www.arm.com, accessed October 2008.

[Ash06] P. Ashenden. *The Designer's Guide to VHDL*. Morgan Kaufmann, 3rd revised edition, 2006.

[ASU03] A.V. Aho, R. Sethi, and J.D. Ullman. *Compilers*. Addison-Wesley Longman, 2003.

[Bra83] D. Brand. Redundancy and Don't Cares in Logic Synthesis. *IEEE Transactions on Computers*, 32(10):947–952, 1983.

[BB+07] J. Bormann, S. Beyer, A. Maggiore, M. Siegel, S. Skalberg, T. Blackmore, and F. Bruno. Complete Formal Verification of TriCore2 and Other Processors. In *DVCon*, 2007.

[BDBK08] D. C. Black, J. Donovan, B. Bunton, and A. Keist. *SystemC: From the Ground Up*. Spinger, 2nd edition, 2008.

[BFF05] N. Bombieri, A. Fedeli, and F. Fummi. On PSL Properties Re-use in SoC Design Flow Based on Transaction Level Modeling. In *International Workshop on Microprocessor Test and Verification*, pp. 127–132, 2005.

[BHJM07] D. Beyer, T.A. Henzinger, R. Jhala, and R. Majumdar. The Software Model Checker Blast: Applications to Software Engineering. *International Journal on Software Tools for Technology Transfer*, 9(5–6):505–525, 2007.

[BKS08] N. Blanc, D. Kroening, and N. Sharygina. Scoot: A Tool for the Analysis of SystemC Models. In *International Conference on Tools and Algorithms for the Construction and Analysis of Systems*, pp. 467–470, 2008.

[BMP07] B. Bailey, G. Martin, and A. Piziali. *ESL Design and Verification: A Prescription for Electronic System Level Methodology*. Morgan Kaufmann, 2007.

[BNL05] D. Beyer, A. Noack, and C. Lewerentz. Efficient Relational Calculation for Software Analysis. *IEEE Transactions on Software Engineering*, 31(2):137–149, 2005.

[BP+05] D. Berner, H. Patel, D. Mathaikutty, J.-P. Talpin, and S. Shukla. SystemCXML: An Extensible SystemC Front End Using XML. In *Forum on Specification and Design Languages*, pp. 405–408, 2005.

[BR02] T. Ball and S.K. Rajamani. The SLAM Project: Debugging System Software via Static Analysis. In *Symposium on Principles of Programming Languages*, pp. 1–3, 2002.

[Cir04] Collett International Research. *IC/ASIC Functional Verification Study*, 2002/2004.

[CoWare] CoWare Inc. Platform Architect Homepage. [Online], http://www.coware.com, accessed July 2008.

[CD+05] Xi Chen, A. Davare, H. Hsieh, A. Sangiovanni-Vincentelli, and Y. Watanabe. Simulation Based Deadlock Analysis for System Level Designs. In *Design Automation Conference*, pp. 260–265, 2005.

[CDT] Eclipse CDT Project Homepage. [Online], http://www.eclipse.org/cdt, accessed September 2008.

[CF+93] J. Cuny, G. Forman, A. Hough, J. Kundu, C. Lin, L. Snyder, and D. Stemple. The Ariadne Debugger: Scalable Application of Event-Based Abstraction. In *Workshop on Parallel & Distributed Debugging*, pp. 85–95, 1993.

[CGP00] E.M. Clarke, O. Grumberg, and D.A. Peled. *Model checking*. MIT Press, Cambridge, MA, 2000.

[CRAB01] L. Charest, M. Reid, E. Aboulhamid, and G. Bois. A Methodology for Interfacing Open Source SystemC with a Third party Software. In *Design, Automation and Test in Europe*, pp. 16–20, 2001.

[CS06] C. Csallner and Y. Smaragdakis. Dynamically Discovering likely Interface Invariants. In *International Conference on Software Engineering*, pp. 861–864, 2006.

[CS+06] E. Cheung, P. Satapathy, Vi Pham, H. Hsieh, and Xi Chen. Runtime Deadlock Analysis of SystemC Designs. In *IEEE International High-Level Design Validation and Test Workshop*, pp. 187–194, 2006.

[Das00] M. Das. Static Analysis of Large Programs (Invited Talk) (Abstract only): Some Experiences. In *ACM/SIGPLAN Workshop Partial Evaluation and Semantics-Based Program Manipulation*, p. 1, 2000.

[Det96] D.L. Detlefs. An overview of the extended static checking system. In *Formal Methods in Software Practice*, pp. 1–9, 1996.

[Don92] C. Donnelly. *Bison: The Yacc – Compatible Parser Generator*. Free Software Foundation, 1992.

[DAC99] M. B. Dwyer, G. S. Avrunin, and J. C. Corbett. Patterns in Property Specifications for Finite-state Verification. In *International Conference on Software Engineering*, pp. 411–420, 1999.

[DE88] M. Ducassé and A.-M. Emde. A Review of Automated Debugging Systems: Knowledge, Strategies and Techniques. In *International Conference on Software Engineering*, pp. 162–171, 1988.

[DF04] R. Drechsler and G. Fey. Design Understanding by Automatic Property Generation. In *Workshop on Synthesis And System Integration of Mixed Information Technologies*, pp. 274–281, 2004.

[DG+05] A. Dahan, D. Geist, L. Gluhovsky, D. Pidan, G. Shapir, Y. Wolfsthal, L. Benalycherif, R. Kamidem, and Y. Lahbib. Combining System Level Modeling with Assertion based Verification. In *International Symposium on Quality of Electronic Design*, pp. 310–315, 2005.

[DG02] R. Drechsler and D. Große. Reachability Analysis for Formal Verification of SystemC. In *Euromicro Symposium on Digital System Design*, pp. 337–340, 2002.

[DLS02] M. Das, S. Lerner, and M. Seigle. ESP: Path-Sensitive Program Verification in Polynomial Time. In *Conference on Programming Language Design and Implementation*, pp. 57–68, 2002.

[DSG03] F. Doucet, S. Shukla, and R. Gupta. Introspection in System-Level Language Frameworks: Meta-Level vs. Integrated. In *Design, Automation, and Test in Europe*, pp. 382–387, 2003.

[EA03] D. Engler and K. Ashcraft. RacerX: Effective, Static Detection of Race Conditions and deadlocks. In *ACM SIGOPS Operating Systems Review*, Volume 37, Issue 5, pp. 237–252, 2003.

[EAH05] C. Eibl, C. Albrecht, and R Hagenau. gSysC: A Graphical Front End for SystemC. In *European Conference on Modelling and Simulation*, pp. 257–262, 2005.

[EC+01] D. Engler, D.Y. Chen, S. Hallem, A. Chou, and B. Chelf. Bugs as Deviant Behavior: A General Approach to Inferring Errors in Systems Code. *ACM SIGOPS Operating Systems Review*, Volume 35, Issue 5, pp. 57–72, 2001

[Fos06] H. Foster. Seven Habits of Effective Formal Verification Planning. In *SOCcentral*, [Online], http://www.soccentral.com, accessed May 2008.

[FD04] G. Fey and R. Drechsler. Improving Simulation-Based Verification by Means of Formal Methods. In *Asia and South Pacific Design Automation Conference*, pp. 640–643, 2004.

[FF01] C. Flanagan and S.N. Freund. Detecting Race Conditions in Large Programs. In *Workshop on Program Analysis for Software Tools and Engineering*, pp. 90–96, 2001.

[FGP07] M. Fujita, I. Gosh, and M. Prasad. *Verification Techniques for System-Level Design*. Morgan Kaufmann Series in Systems on Silicon, 2007.

[FKB08] R. Fechete, G. Kienesberger, and J. Blieberger. A Framework for CFG-Based Static Program Analysis of Ada Programs. In *Ada-Europe International Conference on Reliable Software Technologies*, pp. 130–143, June 2008.

[FKL03] H. Foster, A. Krolnik, and D. Lacey. *Assertion-Based Design*. Kluwer, 2003.

[FTA02] J.S. Foster, T. Terauchi, and A. Aiken. Flow-Sensitive Type Qualifiers. In *Conference on Programming Language Design and Implementation*, pp. 1–12, 2002.

[Ghe06] F. Ghenassia. *Transaction-Level Modeling with SystemC*. Kluwer Academic Publishers, 2006.

[God97] P. Godefroid. Model checking for programming languages using VeriSoft. In *Symposium on Principles of Programming Languages*, pp. 174–186, 1997.

[GCOV] GNU test coverage program gcov. [Online], http://gcc.gnu.org/onlinedocs/gcc/Gcov.html, accessed June 2008.

[GD03] D. Große and R. Drechsler. Formal Verification of LTL Formulas for SystemC Designs. In *IEEE International Symposium on Circuits and Systems*, pp. 245–248, 2003.

[GD04a] D. Große and R. Drechsler. Checkers for SystemC Designs. In *International Conference on Formal Methods and Models for Codesign*, pp. 171–178, 2004.

[GD04b] G. Fey and R. Drechsler. Improving Simulation-Based Verification by Means of Formal Methods. In *Asia and South Pacific Design Automation Conference*, pp. 640–643, 2004.

[GDB] R. Stallman, R. Pesch, St. Shebs, et al. *Debugging with GDB*. Free Software Foundation, Inc., [Online], http://www.gnu.org/software/gdb, accessed July 2008.

[GDLA03] D. Große, R. Drechsler, L. Linhard, and G. Angst. Efficient Automatic Visualization of SystemC Designs. In *Forum on specification and Design Languages*, pp. 646–657, 2003.

[GDPG01] A. Gerstlauer, R. Dömer, J. Peng, and D.D. Gajski. *System Design – A Practical Guide with SpecC*. Springer, 2001.

[GED07] D. Große, R. Ebendt, and R. Drechsler. Improvements for Constraint Solving in the SystemC Verification Library. In *Great Lakes Symposium on VLSI*, pp. 493–496, 2007.

[GHJV95] E. Gamma, R. Helm, R. Johnson, and J. Vlissides. *Design Patterns: Elements of Reusable Object-Oriented Software*. Addison-Wesley, 1995.

[GPKD08] D. Große, H. Peraza, W. Klingauf, and R. Drechsler. Measuring the Quality of a SystemC Testbench by Using Code Coverage Techniques. In *Embedded Systems Specification and Design Languages: Selected Contributions from FDL'07*, pp. 73–86, Springer, 2008.

[GV+94] D.D. Gajski, F. Vahid, S. Narayan, and J. Gong. *Specification and Design of Embedded Systems*. Prentice Hall, Englewood Cliffs, NJ, 1994.

[GWSD08] D. Große, R. Wille, R. Siegmund, and R. Drechsler. Contradiction Analysis for Constraint-Based Random Simulation. In *Forum on Specification & Design Languages*, pp. 130–135, 2008.

[Hum04] W. S. Humphrey. The quality attitude. In *news@sei 2004 | Number 3*, [Online], http://www.sei.cmu.edu/publications/news-at-sei/columns/watts_new/2004/3/watts-new-2004-3.htm, accessed May 2008.

[HCNC05] S. Hangal, N. Chandra, S. Narayanan, and S. Chakravorty. IODINE: A Tool to Automatically Infer Dynamic Invariants for Hardware Designs. In *Design Automation Conference*, pp. 775–778, 2005.

[HCXE02] S. Hallem, B. Chelf, Y. Xie, and D. Engler. A System and Language for Building System-Specific, Static Analyses. In *Conference on Programming Language Design and Implementation*, pp. 69–82, 2002.

[HFG08] P. Herber, J. Fellmuth, and S. Glesner. Model Checking SystemC Designs Using Timed Automata. In *International Conference on Hardware/Software Codesign and System Synthesis (CODES+ISSS)*, pp. 131–136 , 2008.

[HJM04] T.A. Henzinger, R. Jhala, and R. Majumdar. Race Checking by Context Inference. In *International Conference on Programming Language Design and Implementation*, pp. 1–13, 2004.

[HL02] S. Hangal and M.S. Lam. Tracking down Software Bugs using Automatic Anomaly Detection. In *International Conference on Software Engineering*, pp. 291–301, 2002.

[HR06] M. Holzer and M. Rupp. Static Code Analysis of Functional Descriptions in SystemC. In *IEEE International Workshop on Electronic Design, Test and Applications*, pp. 243–248, 2006.

[HR08] C. Haufe and F. Rogin. Ad-Hoc Translations to Close Verilog Semantics Gap. In *IEEE Workshop on Design and Diagnostics of Electronic Circuits and Systems*, pp. 195–200, 2008.

[HT04] A. Habibi and S. Tahar. On the Extension of SystemC by SystemVerilog Assertions. In *Canadian Conference on Electrical and Computer Engineering*, Volume: 4, pp. 1869–1872, 2004.

[HT05] A. Habibi and S. Tahar. On the Transformation of SystemC to AsmL Using Abstract Interpretation. *Electronic Notes in Theoretical Computer Science*, Vol. 131:39–49, 2005.

[HT06] A. Habibi and S. Tahar. Design and Verification of SystemC Transaction-Level Models. *IEEE Transactions on VLSI Systems*, 14(1):57–68, 2006.

[IB06] B. Isaksen and V. Bertacco. Verification through the Principle of Least Astonishment. In *IEEE/ACM International Conference on Computer-Aided Design*, pp. 860–867, 2006.

[IE+05a] IEEE Computer Society. IEEE Standard for SystemVerilog-Unified Hardware Design, Specification, and Verification Language. 2005.

[IE+05b] IEEE Computer Society. IEEE Standard for Property Specification Language. 2005.

[IE+06] IEEE Computer Society. IEEE Standard SystemC Language Reference Manual. 2006.

[IIS] Fraunhofer IIS. "Modeling SystemC" training class. [Online], http://www.iis.fraunhofer.de/bf/train/kurse/sysc, accessed October 2008.

[ITRS07] International Technology Roadmap for Semiconductors – Design, ITRS Edition 2007. [Online], http://www.itrs.net, accessed January 2009.

[Joh97] R.E. Johnson. Frameworks = (components + patterns). In *Communications of the ACM*, 40(10):39–42, 1997.

[Kog06] T. Kogel. Peripheral Modeling for Platform Driven ESL Design. In *Platform Based Design at the Electronic System Level*, pp. 71–85, Springer, 2006.

[Kro99] T. Kropf. *Introduction to Formal Hardware Verification*. Springer, 1999.

[KL88] B. Korel and J. Laski. Dynamic program slicing. *Information Processing Letters*, 29(3):155–163, 1988.

[KP99] B.W. Kernighan and R. Pike. *The Practice of Programming*. Addison-Wesley, 1999.

[KPKG02] A. Kuehlmann, V. Paruthi, F. Krohm, and M.K. Ganai. Robust Boolean Reasoning for Equivalence Checking and Functional Property Verification. *IEEE Transactions in Computer-Aided Design*, 21(12):1377–1394, 2002.

[KS05] D. Kröning and N. Sharygina. Formal Verification of SystemC by Automatic Hardware/Software Partitioning. In *Conference on Formal Methods and Programming Models for Codesign*, pp. 101–110, 2005.

[KSF99] D. Kranzlmüller, N. Stankovic, and J. Volkert. Debugging Parallel Programs with Visual Patterns. In *IEEE Symposium on Visual Languages*, pp. 180–181, 1999.

[KT07] A. Kasuya and T. Tesfaye. Verification Methodologies in a TLM-to-RTL Design Flow. In *Design Automation Conference*, pp. 199–204, 2007.

[LMB92] J. Levine, T. Mason, and D. Brown. *Lex and Yacc: UNIX Programming Tools (A Nutshell Handbook)*. O'Reilly Media, 1992.

[LW+05] M.S. Lam, J. Whaley, V.B. Livshits, M.C. Martin, D. Avots, M. Carbin, and C. Unkel. Context-Sensitive Program Analysis as Database Queries. In *Symposium on Principles of Database Systems*, pp. 1–12, 2005.

[Mar98] F. Martin. PAG – An Efficient Program Analyzer Generator. *International Journal on Software Tools for Technology Transfer*, 2(1):46–67, 1998.

[Mor98] R. Morgan. *Building an Optimizing Compiler*. Digital Press, 1998.

[Muc97] S. Muchnick. *Advanced Compiler Design and Implementation*. Morgan Kaufmann, 1997.

[MMM05a] M. Moy, F. Maraninchi, and L. Maillet-Contoz. LusSy: A Toolbox for the Analysis of Systems-on-a-Chip at the Transactional Level. In *International Conference on Application of Concurrency to System Design*, pp. 26–35, 2005.

[MMM05b] M. Moy, F. Maraninchi, and L. Maillet-Contoz. PINAPA: An Extraction Tool for SystemC Descriptions of Systems-on-a-Chip. In *ACM International Conference on Embedded Software*, pp. 317–324, 2005.

[MR+01] W. Mueller, J. Ruf, D. Hoffmann, J. Gerlach, T. Kropf, and W. Rosenstiehl. The Simulation Semantics of SystemC. In *Design, Automation and Test in Europe*, pp. 64–70, 2001.

[MS06] G. Misherghi and Z. Su. HDD: Hierarchical Delta Debugging. In *International Conference on Software Engineering*, pp. 142–151, 2006.

[NB05] N. Nagappan and T. Ball. Static Analysis Tools as Early Indicators of Pre-Release Defect Density. In *International Conference on Software Engineering*, pp. 580–586, 2005.

[NH06] B. Niemann and C. Haubelt. Assertion-Based Verification of Transaction Level Models. In *Workshop "Methoden und Beschreibungssprachen zur Modellierung und Verifikation von Schaltungen und Systemen"*, pp. 232–236, 2006.

[NOT06] R. Nieuwenhuis, A. Oliveras, and C. Tinelli. Solving SAT and SAT Modulo Theories: From an Abstract Davis-Putnam-Logemann-Loveland Procedure to DPLL(T). *Journal of the ACM*, 53(6):937–977, 2006.

[NS+02] J. Niere, W. Schäfer, J.P. Wadsack, L. Wendehals, and J. Welsh. Towards Pattern-Based Design Recovery. In *International Conference on Software Engineering*, pp. 338–348, 2002

[NW+04] N. Nagappan, L. Williams, J. Hudepohl, W. Snipes, and M. Vouk. Preliminary Results On Using Static Analysis Tools For Software Inspection. In *International Symposium on Software Reliability Engineering*, pp. 429–439, 2004.

[OHT04] K. Oumalou, A. Habibi, and S. Tahar. Design for verification of a PCI bus in SystemC. In *International Symposium on System-on-Chip*, pp. 201–204,2004.

[OSCI] OSCI. SystemC home page. [Online], http://www.systemc.org, accessed July 2008.

[OVL] Accellera Organization. Accellera Standard OVL V2. [Online], http://www.accellera.org, accessed April 2008.

[Pal03] S. Palnitkar. *Verilog HDL: A Guide to Digital Design and Synthesis*. Prentice Hall, 2nd edition, 2003.

[Parr07] T. Parr. *The Definitive ANTLR Reference: Building Domain-Specific Languages*. Pragmatic Programmers, 2007.

[PC95] C. Potter and T. Cory. *CAST Tools: An Evaluation and Comparison*. Bloor Research Group, 1995.

[PC05] S. Park and S. Chae. A C/C++-Based Functional Verification Framework Using the SystemC Verification Library. In *IEEE International Workshop on Rapid System Prototyping*, pp. 237–239, 2005.

[PE04] J.H. Perkins and M.D. Ernst. Efficient Incremental Algorithms for Dynamic Detection of likely Invariants. In *ACM SIGSOFT Symposium on the Foundations of Software Engineering*, pp. 23–32, 2004.

[PS03] E. Pozniansky and A. Schuster. Efficient On-the-Fly Data Race Detection in Multithreaded C++ Programs. In *Principles and Practice of Parallel Programming*, pp. 179–190, 2003.

[Rapp08] C.W. Rapp. SMC Programmer's Manual. [Online], http://smc.sourceforge.net/SmcManual.htm, accessed June 2008.

[RDR09] F. Rogin, R. Drechsler, and Steffen Rülke. Automatic Debugging of System-on-a-Chip Designs. In *IEEE International SOC Conference*, pp. 333–336, 2009.

[RF04] F. Rogin and E. Fehlauer. FSM-Based Rule Specification Aiming at a Generic Code Analysis Library. In *Work-in-Progess Session at EUROMICRO*, 2004.

[Rog02] F. Rogin. *Konzeption und Prototyp-Implementierung eines Parser-basierten Analysesystems für Spezifikationen in e-Code*. Diplomarbeit, BTU Cottbus, 2002.

[RF+07] F. Rogin, E. Fehlauer, S. Rülke, S. Ohnewald, and T. Berndt. Non-Intrusive High-level SystemC Debugging. In *Advances in Design and Specification Languages for Embedded Systems*, pp. 131–144, Springer, July 2007.

[RFHO07] F. Rogin, E. Fehlauer, C. Haufe, and S. Ohnewald. Debug Patterns for Efficient High-level SystemC Debugging. In *IEEE Workshop on Design and Diagnostics of Electronic Circuits and Systems*, pp. 403–408, 2007.

[RFSH05] F. Rogin, E. Fehlauer, A. Schneider, and J. Haase. Automatische Generierung von Dokumentationen für VHDL-AMS-Modellbibliotheken. In *ASIM Workshop*, 2005, Slides [Online], http://swt.cs.tu-berlin.de/asim-sts-05/folien/rogin.pdf.

[RGDR08] F. Rogin, C. Genz, R. Drechsler, and S. Rülke. An Integrated SystemC Debugging Environment. In *Embedded Systems Specification and Design Languages: Selected papers from FDL 2007*, pp. 59–71, Springer, 2008.

[RH06] V.P. Ranganath, and J. Hatcliff. An Overview of the Indus Framework for Analysis and Slicing of Concurrent Java Software. In *International Workshop on Source Code Analysis and Manipulation*, pp. 3–7, 2006.

[RHKR01] J. Ruf, D.W. Hoffmann, T. Kropf, and W. Rosenstiel. Simulation-Guided Property Checking on Multi-Valued AR-Automata. In *Design, Automation and Test in Europe*, pp. 742–748, 2001.

[RK+08] F. Rogin, T. Klotz, G. Fey, R. Drechsler, and S. Rülke. Automatic Generation of Complex Properties for Hardware Designs. In *Design, Automation, and Test in Europe*, pp. 545–548, 2008.

[RK+09] F. Rogin, T. Klotz, G. Fey, R. Drechsler, and S. Rülke. Advanced Verification by Automatic Property Generation. *IET Computers & Digital Techniques*, 3(4):338–353, 2009.

[Sal03] A. Salem. Formal Semantics of Synchronous SystemC. In *Design, Automation and Test in Europe*, pp. 376–381, 2003.

[Shu02] F. Shull et al. What We Have Learned About Fighting Defects. In *IEEE International Symposium on Software Metrics*, pp. 249–258, 2002.

[Sto00] S.D. Stoller. Model-Checking Multi-threaded Distributed Java Programs. In *International SPIN Workshop on SPIN Model Checking and Software Verification*, pp. 224–244, 2000.

[Summit] Summit Design, Inc. Vista Design Environment for SystemC Homepage. [Online] http://www.sd.com, accessed August 2008.

[Syn] Synopsys Inc. Simple 8 bit micro-controller. VCS simulator example package.

[SB+97] S. Savage, M. Burrows, G. Nelson, P. Sobalvarro, and T. Anderson. Eraser: A Dynamic Data Race Detector for Multithreaded Programs. *ACM Transactions on Computer Systems*, 15(4):391–411, 1997.

[SBR02] A. Siebenborn, O. Bringmann, and W. Rosenstiel. Worst-case Performance Analysis of Parallel, Communicating Software Processes. In *International Workshop on Hardware/Software Codesign*, pp. 37–42, 2002.

[SC+96] S. Shende, J. Cuny, L. Hansen, J. Kundu, S. McLaughry, and O. Wolf. Event and State-Based Debugging in TAU: A Prototype. In *Symposium on Parallel and Distributed Tools*, pp. 21–30, 1996.

[SCV] SystemC Verification Group. SystemC Verification Standard Specification - Version 1.0e. [Online], http://www.systemc.org, accessed November 2008.

[SLU05] O. Spinczyk, D. Lohmann, and M. Urban, AspectC++: an AOP Extension for C++. *Software Developer's Journal*, pp. 68–76, 2005.

[SMA04] K. R.G. da Silva, E. U.K. Melcher, and G. Araujo. An Automatic Testbench Generation tool for a SystemC Functional Verification Methodology. In *Symposium on Integrated Circuits and System Design*, pp. 66–70, 2004.

[SO06] N. Shi and R.A. Olsson. Reverse Engineering of Design Patterns from Java Source Code. In *International Conference on Automated Software Engineering*, pp. 123–134, 2006.

[Tas02] G. Tassey. The Economic Impacts of Inadequate Infrastructure for Software Testing. *National Institute for Standards and Technology*, 2002.

[TH07] J. Teich and C. Haubelt. *Digitale Hardware/Software-Systeme*. 2.erweiterte Auflage, Springer-Verlag Berlin Heidelberg, 2007.

[TLM2] TLM Working Group. OSCI TLM2 User Manual. In *TLM2 documentation package*, [Online], http://www.systemc.org, accessed July 2008.

[Ull89] J.D. Ullmann. *Principles of Database and Knowledge-Base Systems*. Computer Science Press, Rockville, MD, 2nd Edition, 1989.

[Vera] Synopsys Inc. OpenVera. [Online], http://www.open-vera.com, accessed May 2008

[Vok06] M. Vokac. An efficient Tool for Recovering Design Patterns from C++ Code. *Journal of Object Technology*, 5(1):139–157, 2006.

[Wei84] M. Weiser. Program slicing. *IEEE Transactions on Software Engineering*, 10(4):352–357, 1984.

[Wen03] L. Wendehals. Improving Design Pattern Instance Recognition by Dynamic Analysis. In *Workshop Dynamic Analysis*, pp. 29–32, 2003.

[Wil01] E.D. Willink. *Meta-Compilation for C++*. PhD Thesis, Computer Science Research Group, University of Surrey, 2001.

[WD+05] A. Wieferink, M. Doerper, T. Kogel, G. Braun, A. Nohl, R. Leupers, G. Ascheid, and H. Meyr. A System Level Processor/Communication Co-Exploration Methodology for Multi-Processor System-on-Chip Platforms. In *IEE Proceedings: Computers & Digital Techniques*, 152(1):3–11, 2005.

[WM96] R. Wilhelm and D. Maurer. *Übersetzerbau. Theorie, Konstruktion, Generierung*. Springer, 2. überarbeitete und erweiterte Auflage, 1996.

[WN05] W. Weimer and G. Necula. Mining Temporal Specifications for Error Detection. In *International Conference on Tools and Algorithm for the Construction and Analysis of Systems*, pp. 461–476, 2005.

[WOLB92] L. Wos, R. Overbeek, E. Lusk, and J. Boyle. *Automated Reasoning: Introduction and Applications*. McGraw-Hill, 2nd edition, 1992.

[XR08] G. Xu and A. Rountev. AJANA: A General Framework for Source-Code-Level Interprocedural Dataflow Analysis of AspectJ Software. In *Conference on Aspect-Oriented Software Development*, pp. 36–47, 2008.

[YE04] J. Yang and D. Evans. Automatically Inferring Temporal Properties for Program Evolution. In *International Symposium on Software Reliability Engineering*, pp. 340–351, 2004

[YE+06] J. Yang, D. Evans, D. Bhardwaj, T. Bhat, and D. Das. Perracotta: Mining Temporal API Rules from Imperfect Traces. In *International Conference on Software Engineering*, pp. 282–291, 2006.

[YPA06] J. Yuan, C. Pixley, and A. Aziz. *Constraint-Based Verification*. Springer, 2006.

[YRC05] Y. Yu, T. Rodeheffer, and W. Chen. RaceTrack: Efficient Detection of Data Race Conditions via Adaptive Tracking. *ACM SIGOPS Operating Systems Review*, 39(5):221–234, 2005.

[Zel05] A. Zeller. *WHY PROGRAMS FAIL – A Guide to Systematic Debugging*. Morgan Kaufmann, 2005.

[Zel08] A. Zeller. Using Delta Debugging - A short tutorial. [Online], http://www.st.cs.uni-sb.de/dd/ddusage.php3, accessed October 2008.

[ZH02] A. Zeller and R. Hildebrandt. Simplifying and Isolating Failure-inducing Input. *IEEE Transactions on Software Engineering*, 28(2):183–200, 2002.

[ZW+06] J. Zheng, L. Williams, N. Nagappan, W. Snipes, J.P. Hudepohl, and M.A. Vouk, On the Value of Static Analysis for Fault Detection in Software. *IEEE Transactions on Software Engineering*, 32(4):240–253, 2006.

List of Acronyms

The following list defines the acronyms used throughout the book:

ABV	Assertion-Based Verification
API	Application Programming Interface
ASM	Abstract State Machine
BDD	Binary Decision Diagrams
BFM	Bus Functional Model
BMC	Bounded Model Checking
CFG	Control Flow Graph
DB_A	REGATTA analysis database
DB_R	REGATTA report database
DFA	Data Flow Analysis
EBNF	Extended Backus Naur Form
ESL	Electronic System Level
EST	Extended Syntax Tree
FDC	FSM-Described Configurations
FSM	Finite-State Machine
HDL	Hardware Description Language
IP	Intellectual Property
OVL	Open Verification Library
PSL	Property Specification Language
RTL	Register Transfer Level
SCV	SystemC Verification Library

SDL	Specification and Description Language
SIMD	Single Instruction Multiple Data
SoC	System-on-a-Chip
SVA	SystemVerilog Assertions
TLM	Transaction Level Modeling
VCD	Value Change Dump

Index of Symbols

(t_i, δ_n)	activation time point at simulation time t_i and delta cycle δ_n of D
$\perp_i$	ith parse point annotated to a grammar rule in Γ
A	analysis alphabet of Γ
$a_{p_i}[t]$	activity of property p_i at time t
$a_{s_i}[t]$	activity of signal s_i at time t
B	basic block
C	set of property checkers
C_{basic}	set of basic property checkers
$C_{complex}$	set of complex property checkers
c_F	failing test case
c_P	passing test case
Δ	delta (difference) between two process schedules
δ_n	nth delta cycle while simulating D
D	SystemC design
f_A	function annotating data flow information to G
G	control flow graph
Γ	context-free grammar
I	set of method and thread processes instantiated in D
Π	set of all parse points $\perp_i$ for Γ
P	set of properties valid on T
P_{cand}	set of property candidates
p_i	property in P valid on T
rx	regular expression string
S	vector of signals in T
S_{EX}	set of excluded signals from S
s_i	signal from S
$s_i[t]$	value (signal state) of signal s_i at time t
S_M	set of signals from S matching rx
$S_d^X(w)$	dynamic slice for variable v at execution position q and program input x with $w = (v, q, x)$
$S_s^X(w)$	static slice for variable v at execution position q with $w = (v, q)$
T	simulation trace
t	time reference
t_{cyc}	length of simulation trace T
$v[t]$	design state at time t
Ψ	process schedule of a

	concrete simulation run of D
X	set of activation time points of D
x^n	n^{th} activation of an instantiated process $x \in I$ while simulating D
Z	FDC specification

Index

Kolibri 360 GmbH
Ferdinand-Jühlke-Straße 7
99095 Erfurt, Deutschland
produktsicherheit@kolibri360.de